OPTIMIZATION THEORY

MATHEMATICS LECTURE NOTE SERIES

J. Frank Adams	LECTURES ON LIE GROUPS
E. Artin and J. Tate	CLASS FIELD THEORY
Michael Atiyah	K-THEORY
Jacob Barshay	TOPICS IN RING THEORY
Hyman Bass	ALGEBRAIC K-THEORY
Melvyn S. Berger Marion S. Berger	PERSPECTIVES IN NONLINEARITY
Armand Borel	LINEAR ALGEBRA GROUPS
Raoul Bott	LECTURES ON K (X)
Andrew Browder	INTRODUCTION TO FUNCTION ALGEBRAS

Gustave Choquet	LECTURES ON ANALYSIS
	    I. INTEGRATION AND TOPOLOGICAL VECTOR SPACES
	    II. REPRESENTATION THEORY
	    III. INFINITE DIMENSIONAL MEASURES AND PROBLEM SOLUTIONS

Paul J. Cohen	SET THEORY AND THE CONTINUUM HYPOTHESIS
Eldon Dyer	COHOMOLOGY THEORIES
Robert Ellis	LECTURES ON TOPOLOGICAL DYNAMICS
Walter Feit	CHARACTERS OF FINITE GROUPS
John Fogarty	INVARIANT THEORY
William Fulton	ALGEBRAIC CURVES
Marvin J. Greenberg	LECTURES ON ALGEBRAIC TOPOLOGY
Marvin J. Greenberg	LECTURES ON FORMS IN MANY VARIABLES
Robin Hartshorne	FOUNDATIONS OF PROJECTIVE GEOMETRY
J. F. P. Hudson	PIECEWISE LINEAR TOPOLOGY

A Note from the Publisher

This volume was printed directly from a typescript prepared by the author, who takes full responsibility for its content and appearance. The Publisher has not performed his usual functions of reviewing, editing, typesetting, and proofreading the material prior to publication.

The Publisher fully endorses this informal and quick method of publishing lecture notes at a moderate price, and he wishes to thank the author for preparing the material for publication.

OPTIMIZATION THEORY

DAVID RUSSELL

University of Wisconsin

W. A. BENJAMIN, INC.

New York   1970

OPTIMIZATION THEORY

Standard Book Numbers: 8053-8364-6 (C)
8053-8365-4 (P)
Library of Congress Catalog Card Number: 79-135370
Manufactured in the United States of America
12345R 43210

W. A. BENJAMIN, INC.
New York, New York 10016

To my father, George Russell, whose
mathematical curiosity rubbed off on
his son

Preface

These lecture notes have been developed in the process of teaching an under-graduate course in optimization methods at the University of Wisconsin, Madison. Thus they represent an attempt to explain certain parts of optimization theory and related methods. Rather than aiming at any completeness of coverage we have put every effort into making understandable those topics which we do treat.

The students taking this course come equipped with usual first course in calculus plus the rudiments of linear algebra, that is, they can solve systems of linear equations and compute determinants but are not generally conversant with vectors and matrices. In these lecture notes we have included an introduction to those parts of advanced calculus, e.g., open and closed sets, properties of continuous functions, gradients, Hessians, etc., and those parts of linear algebra, e.g., inverses of matrices, bilinear and quadratic forms, eigenvalues and eigenvectors, etc., which are necessary for a mean-ingful presentation to students with a limited background in mathematics. These aux-iliary topics are not presented all at once. They are introduced only when we need them and we make almost immediate use of them after their introduction. In this way we hope to maintain the interest of students primarily interested in applications. Our basic teaching goal is to provide the student with a set of tools useful in solving opti-mization problems together with knowledge of where these tools come from and what their limitations are.

We have tried to avoid the welter of jargon which seems to burden much of the literature on optimization. Believing mathematics to be the appropriate "lingua franca" for the exact sciences we have used standard mathematical terminology throughout. Thus, e.g., one will find no reference to "state" and "decision" variables which have served as an endless source of confusion to many students. In fact, by consistently using vector notation, we have been able to present most of the material without ex-plicit reference to coordinate systems. The response from students has been gratifying.

Students taking our course normally follow it up with a course in linear program-ming. Because of this, and because the methods of that subject are quite special, we have not included linear programming in this text. However, linear programs of modest size can be handled with reasonable efficiency using the methods of Chapter 10. Also not included are certain topics normally covered in graduate nonlinear programming courses, such as duality theory, various forms of the constraint qualification, etc. We do treat the Kuhn-Tucker conditions, under somewhat special assumptions, in Chapter 10. The method of proof immediately leads into a gradient projection techniques for problems with linear inequality constraints.

Madison, Wisconsin
February 8, 1970

D. L. Russell

ix

Acknowledgments

Thanks are due to the students who have taken this course, as it has been developed over a period of three years, for their patience, interest and piercing questions. Mrs. Grace Krewson deserves praise for remarkably excellent service in typing the manuscript. I would like to thank W. A. Benjamin, Inc. for their very fine cooperation in making the publication of these lecture notes possible.

Contents

CHAPTER 0.   INTRODUCTION

There is an apocryphal tale which is told within the "aerospace" industry going something like this. It concerns the Navier-Stokes partial differential equations of fluid flow which approximately describe, e. g. , the flow of air around an airfoil, or wing. It is solemnly alleged that if the Wright borthers had been familiar with these equations in 1903 their craft would never have left the ground. For, they would have reasoned, if fluid flow is really as complicated as this, then even if an airplane can be built, obviously no one will be able to control it!

This tale illustrates well the old tradition of the superiority of the gifted experimenter, unencumbered by the misgivings of the theoretician. And, we must admit, such a

1

philosophy has often served us well and may occasionally do

so in the future.   However, the increasing complexity of our

technology together with the accompanying acceleration of

costs are making this approach ever less rewarding.   Today

no aircraft is flown before its flight dynamics have been

analyzed mathematically, simulated on a computer and in

wind tunnel experiments, and compared with the recorded

behavior of existing aircraft.   The story could be repeated

for any other complicated technological enterprise.   There is

still plenty of experimentation to be done (more than ever,

actually) but analytical procedures guide the work, separat-

ing the promising lead from the crackpot notion.

It is not only in the physical sciences that theoretical

and quantitative methods are an indispensable part of the

modus operandi.   In the social and biological sciences such

as economics, psychology, ecology, genetics, molecular

biochemistry, etc. , the role of mathematical models becomes

ever more important as increasingly complex processes are

studied.

There is a by-product of this tendency toward quanti-

fication which will be our concern in this text.   As soon as a mathematical description of a process is available, showing

how that process responds to varying factors in its environ-

ment, it becomes possible to pose, and sometimes even to

answer, the question, "What combination of these factors

causes the process to operate in the best possible way? " By

"best possible" we mean, of course, most desirable from the

particular viewpoint we have in mind.   Perhaps, e. g. , we

wish to find the combination of tax and interest rates which

cause the national economy to exhibit some combination of

growth and stability which experience indicates to be most

favorable.

It is our purpose in this text to study techniques for

finding the best possible, i. e. , the optimum.   This field of

optimization theory and technique lies in the ever widening

interdisciplinary area which overlaps onto mathematics, com-

puter science, operations research, statistics, engineering,

economics and many others.   Such a diversity of backgrounds

requires a "lingua franca" if there is to be any communication.

That common language is found in mathematics and our treat-

ment of the subject reflects this fact.   Nevertheless, we will

not be taking our mathematics "straight".   It will be combined

with computational considerations and many examples to
illustrate the very wide range of applications.

We will complete this introductory section by discus-
sing some very simple examples which indicate what we mean
by an optimization problem. The solution of these problems
will require nothing in the way of advanced technique. Nevertheless, they exhibit many of the features which we will en-
counter later in more complicated settings.

Let us suppose that we are renting an automobile at
75¢ per hour, the gasoline to be supplied by us at 35¢ per
gallon. We propose to drive from New York to Chicago, a
distance of about 900 miles. Prudence dictates that we travel
at some speed between 40 and 80 miles per hour. We will
assume that, in this speed range, the mileage derived from
our dearly bought fuel is adequately described by

$$m = 26 - s/5$$

where m is the number of miles per gallon and s is the
speed in miles per hour. Our goal is to make the trip as in-
expensively as possible by judicious choice of s .

We proceed as follows. If we travel at s miles per

hour then our rental cost in dollars will be

$$c_1(s) = \text{(hourly rental)} \times \text{(hours)} = (.75)(\frac{900}{s}) = \frac{675}{s} .$$

On the other hand, our fuel cost will be

$$c_2(s) = \text{(cost per gallon)} \times \text{(gallons)} = (.35)(\frac{900}{m})$$

$$= \frac{315}{m} = \frac{315}{26 - s/5} .$$

We want to minimize the total cost

$$c(s) = c_1(s) + c_2(s) = \frac{675}{s} + \frac{315}{26 - s/5} = \frac{180s + 17,550}{26s - s^2/5} .$$

Typical costs are $35.35 at 40 m. p. h. , $33.75 at 60 m. p. h., and $39.95 at 80 m. p. h. To find the optimum speed, i. e. , the speed for which the cost is least, we must find that value of s in the interval $40 \le s \le 80$ for which $c(s)$ is minimal.

We recall from calculus that if the differentiable function $c(s)$ achieves a minimum value at a point s_0 in the open interval $(40, 80)$ then $c'(s_0) = 0$. Now

$$c'(s) = \frac{(26s - s^2/5)180 - (180s + 17,550)(26 - 2s/5)}{(26s - s^2/5)^2}$$

$$= \frac{36s^2 + 7,020s - 456,300}{(26s - s^2/5)^2}$$

is defined everywhere in (40, 80) and vanishes if the numerator vanishes, i. e. , if

$$36s^2 + 7,020s - 456,300 = 0.$$

The roots of this quadratic equation are, approximately

$$s_1 = -247.2, \qquad s_2 = 52.2 .$$

The first of these does not represent a meaningful speed and is disregarded as extraneous. We take $s_0 = s_2 = 52.2$ and calculate that the cost at this speed is \$33.20. Since $c'(s) < 0$ for $40 < s < s_0$ and $c'(s) > 0$ for $s_0 < s < 80$ it is clear that $s_0 = 52.2$ m. p. h. is the unique speed yielding minimum cost.

Probably the question of existence, i. e. , whether or not there is a best speed at which to travel, hardly occurs to anyone when confronted with this problem. Perhaps it is sensed, more or less intuitively, that the cost is greater than some fixed positive amount no matter what the speed in the given range and that, therefore, there must be a least positive cost. This reasoning is correct but the actual proof of its correctness is not entirely trivial.

Not all optimization problems have solutions. Consider for example the problem of designing a disc-shaped "seltzer" pill in such a way as to minimize the time required for it to dissolve in a glass of water. This means, assuming the standard dosage to be one unit volume, that we wish to find a cylinder of unit volume having maximal surface area. Letting the thickness of the cylinder from top to bottom be h and the radius r, the problem becomes

$$\text{maximize} \quad 2\pi r^2 + 2\pi rh$$

subject to the condition

$$\pi r^2 h = 1.$$

Solving for h in terms of r in this last equation and substituting $h = \dfrac{1}{\pi r^2}$ in the expression to be maximized, the problem assumes the form

$$\text{maximize} \quad 2\pi r^2 + \frac{2}{r}, \qquad r > 0 .$$

The expression $2\pi r^2 + \dfrac{2}{r}$ becomes arbitrarily large as r approaches infinity and as r approaches zero, corresponding to the clyinder becoming an infinite two dimensional plane or

an infinite two dimensional plane or an infinite line, respectively.   Thus no finite value of r can yield a maximum for the
surface area.   Mathematically speaking, the extremes $r = \infty$
and r = 0 correspond to the pill having zero volume - not to
mention the inconvenience which would arise in packaging!

We see then that it is easy to pose problems with no
solution.   In the above problem a little reflection at the out-
set would have told us we were asking for the impossible.   In
more complicated cases it may not be clear whether there is
a solution or not.

It is quite possible to pose optimization problems
having more than one solution, in which case we say that the
solution is not unique.   For example, suppose two stores lie
on the same side of a street, 500 feet apart.   We may park
our car anywhere beside the curb but must then visit both
stores before returning to it.   Where should we park so as to
minimize the total walking distance?   The answer, as the
reader can easily verify, is anywhere between the two stores.
For any of these parking positions we must walk 1000 feet.   If
we park anywhere else we will have to walk a greater distance.

CHAPTER 1.   FUNDAMENTAL CONCEPTS

Probably the most basic notion we need in studying any process is that of the _state_ of the process.   For our purposes the state of a process is determined by selecting a number of phenomena associated with the process, which serve to adequately describe what is going on, and assigning real number values to the observed phenomena in some standard way, e. g. , by the use of accepted units for physical measurements.   If our process consists of a mixture of hydrogen and oxygen gases in a pressurized container, then for some purposes an adequate specification of the process state might consist of four real numbers H, O, T, P, where H is the number of moles of hydrogen present, O the number of moles of oxygen, T the temperature in degrees centrigrade and P the pressure in grams per square centimeter. If our process

is the operation of a warehouse storing bulk commodities and

if n such commodities are normally in inventory, then the

process state might be adequately specified by $c_1, c_2, ..., c_n$,

where c_i is the number of tons of the i-th commodity current-

ly on hand.

In general we will study processes whose states are

describable by an ordered n-tuple of real numbers,

$$x = \begin{pmatrix} x^1 \\ x^2 \\ \vdots \\ x^n \end{pmatrix}$$

Such an n-tuple of real numbers is called an n-dimen-

sional vector or simply an n-vector. The collection of all

such vectors is the real n-dimensional vector space and will

be denoted by the symbol R^n. The real numbers $x^1, x^2, ..., x^n$

are called the components of the vector x. We distinguish

components of the same vector by superscripts. We distin-

guish different vectors by different letters or by the same

letter with different subscripts. This convention is particu-

larly useful when we wish to refer to a sequence of vectors:

$\{x_k\} = \{x_1, x_2, x_3, \ldots\}$.   For each vector x_k we have

$$x_k = \begin{pmatrix} x_k^1 \\ x_k^2 \\ \vdots \\ x_k^n \end{pmatrix}$$

which should help to clarify the role of superscript and sub-

script. The only place where this notation might cause us

some difficulty is a situation where we wish to raise some

component of a vector to a power. We will take care of this

contingency with parentheses. Thus the m-th power of the

j-th component of the vector x is denoted by $(x^j)^m$.

    In R^n we recognize two basic operations on vectors –

addition of vectors and multiplication of vectors by real num-

bers. Both operations are carried out componentwise. Thus

if x and y are vectors their <u>sum</u> is the vector z with

$$z^j = x^j + y^j, \qquad j = 1, 2, \ldots, n.$$

For example, in R^3 the sum of the vectors $\begin{pmatrix} 3 \\ -7 \\ 5 \end{pmatrix}$ and $\begin{pmatrix} -2 \\ 3 \\ 4 \end{pmatrix}$

is $\begin{pmatrix} 1 \\ -4 \\ 9 \end{pmatrix}$.   If α is a real number and x an n-vector, αx

is the vector z with

$$z^j = \alpha x^j, \quad j = 1, 2, \ldots, n .$$

Thus $3 \begin{pmatrix} -2 \\ 3 \\ 4 \end{pmatrix} = \begin{pmatrix} -6 \\ 9 \\ 12 \end{pmatrix}$. The following definitions and

rules apply to these operations.

(i) $-x = (-1)x$;

(ii) $0x = 0 = \begin{pmatrix} 0 \\ 0 \\ \vdots \\ 0 \end{pmatrix}$, the zero vector;

(iii) $\alpha(x + y) = \alpha x + \alpha y$;

(iv) $(\alpha + \beta)x = \alpha x + \beta x$.

Another notion which we will use a great deal is that

of a <u>set</u> of states, i. e. , a set of vectors. This is just some

collection of vectors in R^n which we can describe in some

way. For example, we might describe a set S of vectors in

R^n by saying that S consists of all x such that x^j is pos-

itive for j = 1, 2, ..., n. Or again, we might define a set

T of vectors in R^n by agreeing that x is in R^n if

$\sum_{j=1}^{n} (x^j)^2 \leq 1.$ S is the positive orthant in R^n while T is

the closed unit ball in R^n.

We use the symbol $x \in S$ to indicate that the vector

x belongs to the set S. If we have a number of sets, S, T, U, etc. , all consisting of vectors in the same vector space R^n, we can form further sets by applying certain basic operations to these original sets. These basic operations are $\cup$ (union), $\cap$ (intersection), and ^{C} (complementation). The union of S and T, S $\cup$ T, is the set of all vectors x in R^n such that x ϵ S or x ϵ T. The intersection, S $\cap$ T, is the set of all x such that x ϵ S and x ϵ T. The complement of S, S^C, is the set of all x such that x does not belong to S (symbol x $\notin$ S). From these basic operations it is possible to construct other operations, e. g. , the difference S - T is defined by

$$S - T = S \cap T^C. \quad \text{(The operation} \quad ^C \quad \text{is performed first.)}$$

In order that these operations on sets should always be defined it is necessary to introduce the empty set, denoted ϕ, defined as the set containing no vectors. Thus, if S and T have no members in common, we have

$$S \cap T = \phi,$$

in which case we say S and T are disjoint. If S is any

set, then

$$S \cup \phi = S, \quad S \cap \phi = \phi, \quad S - \phi = S \cap \phi^C = S \cap R^n = S.$$

There is a very convenient notation for describing sets known as the set builder. It consists of two braces and a vertical line segment, or bar. Between the first brace and the bar is a general description of the sort of elements of which the set consists. Between the bar and the second brace is a statement which specifies exactly which of these elements do in fact belong to this particular set. For example, the positive orthant and the closed unit ball are, respectively, described by

$$S = \{x \in R^n \mid x^j > 0, \quad j = 1, 2, \ldots, n\},$$

$$T = \{x \in R^n \mid \sum_{j=1}^{n} (x^j)^2 \leq 1\}.$$

If each vector x which belongs to S also belongs to T, we say that S is included in T, or that S is a subset of T. We write

$$S \subseteq T \quad \text{or} \quad T \supseteq S.$$

If $S \subseteq T$ but $S \neq T$, so that there is at least one element of

T which does not belong to S, then we say that S is <u>pro-</u>
<u>perly</u> <u>included</u> in T or that S is a <u>proper</u> <u>subset</u> of T. We
easily see that for any two sets S and T of vectors in R^n
we have

$$S \cap T \subseteq S \subseteq S \cup T, \quad S - T \subseteq S.$$

When posing an optimization problem we assign a
<u>cost</u> to each possible state of the process in question. This
is a real number uniquely specified in some well defined manner by the given state. For example, in the case of the warehouse, perhaps it costs α_i dollars per day to store one ton
of the i-th commodity. Then the state $c_1, c_2, \ldots, c_n$ gives
rise to a cost $\alpha_1 c_1 + \alpha_2 c_2 + \ldots + \alpha_n c_n$.

Costs of this type are particular cases of functions.
In general we define a <u>function of n variables</u> to be a correspondence f which assigns to each vector x in some subset
D of R^n a unique real number f(x), called the value of f
at x. D is called the <u>domain</u> of f. The set of real numbers

$$R = \{f(x) \in R^1 \mid x \in D\}$$

is called the <u>range</u> of f. We will also have occasion to consider <u>vector</u> <u>valued</u> <u>functions.</u>   These are defined in the

same way.   An m-vector valued function of an n-vector vari-

able is a correspondence h assigning to each n-vector

$x \in D \subseteq R^n$ a uniquely determined m-vector $h(x) \in R^m$.   Again

D is the domain of h and the set

$$\{h(x) \in R^m \mid x \in D\}$$

is the range of h.   We will use the symbol

$$h : R^n \to R^m \quad \text{(read: h maps } R^n \text{ into } R^m\text{)}$$

to designate such a function defined on some domain D in R^n.

It is convenient to use the symbol h(S) to stand for

the set

$$h(S) = \{y \in R^m \mid y = h(x) \quad \text{for some } x \in S \cap D\}.$$

The basic ingredients of an optimization problem are

(i) a set $S \subseteq R^n$, representing all states of the process under

consideration, and (ii) a function $f : R^n \to R^1$ whose domain

includes S, which we may interpret as a cost function or pay-

off function, depending upon our point of view.   We pose an

optimization problem by asking one or more of the following

questions.

(a) Does there exist a vector $x^*\in S$ with the property that $f(x^*) \leq f(x)$ for all $x\in S$? If so, x^* is called a (global) minimum for f in S and $f(x^*)$ is called the minimum value of f in S. If the inequality is reversed we speak of a (global) maximum and a maximum value.

(b) If the answer to (a) is "yes", is there only one such point x^*?

(c) If the answer to both (a) and (b) is "yes", how may we find x^*, or, if this is impossible, how may we find vectors lying arbitrarily near to x^*? We may also ask this question if the answer to (b) is "no" but then we must be content with any x^* unless there are other criteria singling out one of them as particularly desirable.

Question (a) is the question of the existence of solutions. Question (b) is that of the uniqueness of solutions. Both of these are important but the practically oriented person is normally much more interested in (c) which demands an analytical or computational technique producing x^* or some vector near x^*. The bulk of this text will be devoted to (c)

but we ask the reader to be patient while we first discuss (a)
and (b), gently reminding him that (c) may not make sense
unless we have satisfactory answers for (a) and (b), as indicated by the pill problem in the introduction.

A slight modification of (a) is

(d) Do there exist points x^* in S such that $f(x^*) \leq f(x)$
whenever x is a point in S lying near x^*? If so, x^* is
called a <u>local</u> <u>minimum</u> of f in S and $f(x^*)$ is a <u>local</u>
<u>minimum</u> <u>value.</u> For example, the function $f(x) = x^2 - x^4$
has a local minimum at x = 0 but has no global minimum. It
should be observed that global minima are also local minima.

Questions (c) and (d) both involve the concept of
"nearness" of one vector to another. We must now be more
specific concerning what we mean by the word "near".

For $x \in R^n$ we define a <u>norm</u> of x, written $\|x\|$, to
be a function which has the following properties:

(i) $\|x\| \geq 0$ and $\|x\| = 0$ if and only if x = 0;

(ii) $\|ax\| = |a| \|x\|$ for each real number a;

(iii) $\|x+y\| \leq \|x\| + \|y\|$, (Minkowski inequality).

Three norms are commonly used in R^n.

The <u>Euclidean</u> <u>norm</u> is defined by

$$\|x\|_e = (\sum_{j=1}^{n} (x^j)^2)^{\frac{1}{2}}$$

and is usually most convenient for theoretical work. It corresponds with our intuitive concepts of length and distance. The supremum norm ("sup" norm) is

$$\|x\|_s = \max_{j=1,\dots,n} \{|x^j|\},$$

the largest absolute value of any component of x. The "taxicab" norm is given by

$$\|x\|_t = \sum_{j=1}^{n} |x^j|.$$

The norms $\|x\|_s$ and $\|x\|_t$ are more convenient for computation than is the Euclidean norm $\|x\|_e$. Whenever the symbol $\|x\|$ is used in this text it will be understood that what is said applies to all of these norms unless we specifically indicate otherwise.

Given a norm $\|x\|$ defined on R^n we can define what we mean by distance. The distance between two vectors x and y is defined to be the number $\|x - y\|$. Corresponding to the three norms defined above there are three distances.

For example, if $x = \begin{pmatrix} x^1 \\ y^1 \end{pmatrix}$ and $y = \begin{pmatrix} x^2 \\ y^2 \end{pmatrix}$ are vectors in

R^2 as shown in Fig. 1, we construct a right triangle with

sides a and b parallel to the coordinate axes and hypote-

nuse c joining x and y. Then the Euclidean distance

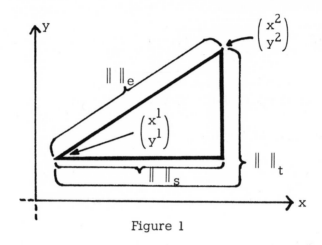

Figure 1

from x to y is the length of c. The "sup" distance is the

length of the longer of a and b. The "taxicab" distance is

obtained by "driving around the block", i.e. it is the length

of a plus the length of b.

All of these distances obey the following rules.

(i) $\|x-y\| \geq 0$ and $\|x-y\| = 0$ if and only if $x = y$.

(ii) $\|x-y\| = \|y-x\|$.

(iii) $\|x-y\| \leq \|x-z\| + \|z-y\|$ for any $z \in R^n$.
(Triangle inequality.)

With our ideas of "nearness" made more precise by the introduction of these distances, we can improve upon the statement of questions (c) and (d)

(c') If the answer to both (a) and (b) is "yes", how may we find x^*, or, if this is impossible, how may we find a sequence of vectors $\{x_k\}$ with the property that

$$\lim_{k \to \infty} \|x^* - x_k\| = 0?$$

(d') Do there exist points x^* in S such that, for some $\delta > 0$, $f(x^*) \leq f(x)$ whenever x is a point in S satisfying

$$\|x^* - x\| < \delta?$$

As stated earlier, the point x^* described in (d') is called a <u>local</u> <u>minimum</u> of f in S, as contrasted with a point x^* described by (a), which is a <u>minimum</u>, or, for emphasis, <u>global</u> <u>minimum</u>, of f in S. For example, using again the function $f(x) = x^2 - x^4$, the point $x = 0$ is a local minimum because

$$f(x) < 0 = f(0) \quad \text{for} \quad |0 - x| = |x| < 1.$$

The points $\pm 1/\sqrt{2}$ are global maxima. We have $f(\pm 1/\sqrt{2}) = \frac{1}{4}$

and for <u>all</u> x

$$f(x) = x^2 - x^4 = \tfrac{1}{4} - (\tfrac{1}{4} - x^2 + x^4) = \tfrac{1}{4} - (x^2 - \tfrac{1}{2})^2.$$

As a rule, approximation methods which provide a
sequence $\{x_k\}$ as described in (d') actually yield sequences
which converge to a local minimum or local maximum, as the
case may be. This may or may not be a global minimum.
There are important exceptions to this rule which will be
pointed out later.

In closing this chapter we remark that whether we
study maximization or minimization is immaterial. For if x^*
is a maximum of f it is a minimum of -f, and vice versa.
Perhaps it depends upon one's outlook - the optimist wants
to maximize his profits, the pessimist wants to minimize
his losses. Without resigning ourselves to pessimism, we
will state most of our problems as minimization problems
simply for convenience. In a few cases maximization problems are easier to talk about and we treat them.

EXERCISES. CHAPTER 1

1. Take n = 2 and sketch the following sets in R^2:

(i) $\{x \in R^2 \mid 1 \leq \|x\|_e \leq 2\}$ (ii) $\{x \in R^2 \mid \|x\|_s \leq 1\}$

(iii) $\{x \in R^2 \mid \|x\|_t \leq 1\}$ (iv) $\{x \in R^2 \mid \|x\|_e \leq 1, \|x\|_t \geq 1\}$

(v) $\{x \in R^2 \mid \|x-y\|_s \leq 1$ for some y with $\|y\|_e = 2\}$.

2. Prove that for arbitrary sets A, B, and C in R^n:

(i) $A \cap B = (A^C \cup B^C)^C$

(ii) $(A - B) \cup (B - A) = ((A \cap B) \cup (A^C \cap B^C))^C$

(iii) $A \cap (B \cap C) = (A \cap B) \cap C$

$= ((A \cap B) \cup (A \cap C) \cup (B \cap C)) - (((A \cap B) - C) \cup ((A \cap C) - B)$

$\cup ((B \cap C) - A))$.

3. Show that there are positive numbers m, M, which do not depend upon x, such that for all $x \in R^n$

$$m \|x\|_s \leq \|x\|_e \leq M \|x\|_s .$$

Repeat the exercise with the pairs $\|x\|_t$, $\|x\|_s$ and $\|x\|_t$, $\|x\|_e$. It is in this sense that these three norms are <u>equivalent.</u>

4. Give examples of four functions $f : R^1 \to R^1$ which have (one at a time) the following:

(i)   a unique global maximum;

(ii) nonunique global minima;

(iii) local maxima and local minima but no global maxima

        and no global minima;

(iv) no local maxima or local minima.

5. Identify global and local maxima and minima of the given

     functions f in the prescribed sets S:

   (i) $f(x) = \|x\|_e$, $S = \{x \in R^2 \mid \|x\|_s \leq 1\}$;

   (ii) $f(x, y) = x^2 - y^2$, $S = \{(\begin{smallmatrix} x \\ y \end{smallmatrix}) \in R^2 \mid x^2 + y^2 \leq 1\}$;

   (iii) $f(x) = -x^4 + x^2 - \tfrac{1}{4}x$, $S = R^1$;

   (iv) $f(x) = x \sin x$, $S = R^1$;

   (v) $f(x, y) = e^{x+y^2}$, $S = \{(\begin{smallmatrix} x \\ y \end{smallmatrix}) \in R^2 \mid x \geq - |y|\}$.

CHAPTER 2   EXISTENCE THEORY

In this chapter we will study the question of the existence of minima and maxima of functions within prescribed sets.
We are given a set $S \subseteq R^n$ and a real valued function $f: R^n \rightarrow R^1$ whose domain D includes S. We ask: does there exist
at least one vector $x^* \in S$ such that

$$f(x) \geq f(x^*), \qquad x \in S ?$$

That is, does f possess a minimum in S? We pose the maximization problem similarly.

The example of the "seltzer" pill cited in the introduction shows that the answer may well be "no". In that exam-
ple we have $f(r) = 2\pi r^2 + \dfrac{2}{r}$ defined on the set $S \subseteq R^1$ given
by $S = \{r \in R^1 | r > 0\}$. Let us change the problem now by

adding the restriction that the radius r of the pill should not

exceed one unit nor be less than one half unit. Thus r is to

lie in the set $S_1 = \{r \in R^1 | \frac{1}{2} \le r \le 1\}$ and our problem becomes

$$\text{maximize } 2\pi r^2 + \frac{2}{r}, \qquad r \in S_1.$$

We solve this problem by noting that $f'(r) = 4\pi r - \dfrac{2}{r^2}$

vanishes at $r_0 = (\frac{1}{2\pi})^{1/3}$, is negative for $r < r_0$ and posi-

tive for $r > r_0$. Consequently $f(\frac{1}{2}) > f(r) > f(r_0)$, $\frac{1}{2} < r < r_0$,

$f(1) > f(r) > f(r_0)$, $r_0 < r < 1$, and it follows that the larger

of $f(\frac{1}{2})$ and $f(1)$ is the desired maximum. Now

$$f(\tfrac{1}{2}) = 2\pi (\tfrac{1}{2})^2 + \frac{2}{\frac{1}{2}} = \frac{2\pi}{4} + r = 5.57$$

$$f(1) = 2\pi (1)^2 + \frac{2}{1} = 2\pi + 2 = 8.28$$

and we have shown that $f(r) \le f(1)$, $r \in S_1$.

We see, therefore, that while the original problem had

no solution, the new problem does. Now we did not change

the function f. All we changed was the set S, replacing it

by S_1 . We ask, what properties does S_1 enjoy that are not

posessed by S? We list two of them:

(i) S_1 is a "bounded" set; all of its points lie within

a unit distance of the origin r = 0. This is not true

of S.

(ii) S_1 includes all of its endpoints, namely $r = \frac{1}{2}$ and $r = 1$. S does not include its left hand endpoint $r = 0$ and has no right hand endpoint.

Note that the failure of S to include its endpoints is directly related to the difficulty experienced with the original problem. For the function f approaches infinity as r tends to 0 and as r tends to $+\infty$ but never is brought to account for this bad behavior since neither 0 nor $+\infty$ represent points of S.

The resolution of the question of the existence of maxima or minima of a function f defined on a set S depends upon properties of f and upon whether or not S has properties similar to (i) and (ii) above. In order to make these ideas precise we need a number of definitions.

DEFINITION 2.1. Let $y \in R^n$. The <u>open ball of radius</u> $\underline{\delta}$ about y is the set

$$N(y, \delta) = \{x \in R^n \mid \|y - x\| < \delta\}, \qquad \delta > 0.$$

Such a ball has a different shape depending upon which norm

we work with.   See Exercise 1 at the end of Chapter 1.

DEFINITION 2. 2.   A set $N \subseteq R^n$ is a <u>neighborhood</u> of the
point (i. e. , vector) y if N includes the open ball $N(y, \delta)$
for some $\delta > 0$.

This definition seems rather generous since it causes
the whole space R^n to be a neighborhood of every point in
R^n. If one wishes to have the term "neighborhood of y" correspond to points lying "near" y then one should think in
terms of balls $N(y, \delta)$ centered at y. We will call these
<u>spherical neighborhoods</u> of y.   Thus a set N is a neighbor-
hood of y if it includes a spherical neighborhood of y.
Exercise 1 of Chapter 1 shows that this definition of neighborhood does not depend upon which norm we choose to use.

DEFINITION 2. 3.   A set $S \subseteq R^n$ is <u>open</u> if it is a neighborhood
of each of its points, i. e., if for each $y \in S$, $N(y, S) \subseteq S$ for
some $\delta > 0$.

It is clear that the whole space R^n is an open set. A
slightly less trivial example is this: we show that an open
ball $N(y, \delta)$ is an open set in the sense of Definition 2. 3.

Let x be a point in $N(y, \delta)$. Then $\|y - x\| < \delta$ so there is a positive number ρ such that $\|y - x\| = \delta - \rho$. We claim that $N(x, \rho) \subseteq N(y, \delta)$. For if $w \in N(x, \rho)$, by the triangle inequality we have

$$\|y - x\| \leq \|y - x\| + \|x - w\| < (\delta - \rho) + \rho = \delta$$

and hence $w \in N(y, \delta)$. Thus $N(y, \delta)$ is open.

DEFINITION 2. 4. A point $y \in R^n$ is a <u>boundary point</u> of a set $S \subseteq R^n$ if every neighborhood of y includes points of S and points of S^c. The set of all boundary points of S is called the <u>boundary</u> of S and is denoted by the symbol ∂S.

Consider again the set $N(y, \delta)$ and let x be such that $\|y - x\| = \delta$. Let N be any neighborhood of x. Then $N(x, \rho) \subseteq N$ for some $\rho > 0$. We will show that $N(x, \rho)$ includes points of $N(y, \delta)$ and points of $N(y, \delta)^c$ and thus is a boundary point of $N(y, \delta)$. We may assume that $\rho < \delta$ for if it were not we could replace it by a number $\rho' < \delta$ and we would still have $N(x, \rho') \subseteq N$. Let $x_1 = (1 - \frac{\rho}{2\delta})x + \frac{\rho}{2\delta}y$, $x_2 = (1 + \frac{\rho}{2\delta})x - \frac{\rho}{2\delta}y$. Then

$$\|y - x_1\| = \|y - (1 - \frac{\rho}{2\delta})x - \frac{\rho}{2\delta}y\| = \|(1 - \frac{\rho}{2\delta})(y - x)\|$$

$$= |1 - \frac{\rho}{2\delta}| \|y - x\| = (1 - \frac{\rho}{2\delta})\delta < \delta,$$

$$\|x - x_1\| = \|x - (1 - \frac{\rho}{2\delta})x - \frac{\rho}{2\delta}y\| = \|\frac{\rho}{2\delta}(x - y)\|$$

$$= |\frac{\rho}{2\delta}| \|x - y\| = (\frac{\rho}{2\delta})\delta = \frac{\rho}{2} < \rho$$

and thus x_1 is a point of $N(x, \rho)$ which is also a point of $N(y, \delta)$ so that N includes points of $N(y, \delta)$. On the other hand

$$\|y - x_2\| = \|y - (1 + \frac{\rho}{2\delta})x + \frac{\rho}{2\delta}y\| = \|(1 + \frac{\rho}{2\delta})(x - y)\|$$

$$= |1 + \frac{\rho}{2\delta}| \|x - y\| = (1 + \frac{\rho}{2\delta})\delta > \delta$$

$$\|x - x_2\| = \|x - (1 + \frac{\rho}{2\delta})x + \frac{\rho}{2\delta}y\| = \|\frac{\rho}{2\delta}(y - x)\|$$

$$= |\frac{\rho}{2\delta}| \|y - x\| = (\frac{\rho}{2\delta})\delta = \frac{\rho}{2} < \rho$$

so that x_2 is a point of $N(x, \rho)$ which is not a point of $N(y, \delta)$ and thus N includes points of $N(y, \delta)^C$. Therefore if $\|y - x\| = \delta$, x is a boundary point of $N(y, \delta)$. It is easy to see that all boundary points x of $N(y, \delta)$ satisfy $\|y - x\| = \delta$ so that

$$\partial N(y, \delta) = \{x \in R^n | \|y - x\| = \delta\}.$$

DEFINITION 2. 5.   A set $S \subseteq R^n$ is <u>closed</u> if it includes all

of its boundary points, i. e. , if $\partial S \subseteq S$.

The set

$$\overline{N(y, \delta)} = \{x \in R^n \mid \|y - x\| \leq \delta\}$$

is closed.   Just as above we can verify that

$$\partial \overline{N(y, \delta)} = \partial N(y, \delta) = \{x \in R^n \mid \|y - x\| = \delta\}$$

and $\overline{N(y, \delta)}$ includes these points.   Perhaps more surprising

is the fact that the whole space R^n is closed.   The whole

space R^n has no boundary points, i. e. , $\partial R^n = \phi$.   The empty

set ϕ is considered to be a subset of every set so

$$\partial R^n = \phi \subseteq R^n$$

and R^n is closed.   Thus R^n furnishes us with an example

of a set which is both open and closed.

Some sets are neither open nor closed.   An example is

the interval $0 \leq x < 1$ in R^1 which includes the boundary

point 0 but not the boundary point 1.   Since it does not in-

clude one of its boundary points, it is not closed.   Also this

set does not include any spherical neighborhood of 0 (i. e. , any open interval centered at 0) and consequently is not open.

THEOREM 2.1. A set $S \subseteq R^n$ is open if and only if S^C is closed.

Proof. We note from Definition 2.4 that if y is a boundary point of S then it must also be a boundary point of S^C. For every neighborhood N of y includes points of S^C and points of $(S^C)^C = S$. Similarly, every boundary point of S^C is also a boundary point of $S = (S^C)^C$. It is also clear that an open set S can include none of its boundary points. For if $y \in S$ there is a neighborhood of y which is included in S, namely $N(y, \delta)$ for some $\delta > 0$, and such a neighborhood of y includes no points of S^C.

Now suppose S is open. Let y be a boundary point of S^C. Then $y \in \partial S$ also and, since S is open, $y \notin S$. But then $y \in S^C$. Thus $\partial S = \partial S^C \subseteq S^C$ and S^C is closed.

Conversely, suppose S^C is closed. Then $\partial S = \partial S^C \subseteq S^C$ and S contains none of its boundary points. Thus if $y \in S$ there must be a neighborhood N of y containing only points of S or only points of S^C. Since $y \in S$ and $y \in N$, N must contain only points of S. Thus there is a neighborhood of y which is contained in S.   Since this is true of any

point y ∈ S, S must be open.   This completes the proof.

For an arbitrary set $S \subseteq R^n$, open, closed or neither,

we can define certain important sets S^o and $\overline{S}$ which are

open and closed respectively.   The definitions are

$$S^o = S - \partial S, \quad \overline{S} = S \cup \partial S.$$

S^o is called the _interior_ of S while $\overline{S}$ is called the _closure_

of S. S^o includes only those points y of S for which

there is a neighborhood of y which is included in S.   We

have the relationships

$$(S^C)^o = (\overline{S})^C, \quad (\overline{S^C}) = (S^o)^C$$

which are easily verified.   One should not confuse the con-

cept of interior with the notion of "inside".   The circle

$\{\begin{pmatrix} x \\ y \end{pmatrix} \in R^2 \mid x^2 + y^2 = 1\}$ consists entirely of boundary points

of itself and thus has an empty interior.   In particular the

interior does not include the points "inside" the circle, i. e.

the points $\begin{pmatrix} x \\ y \end{pmatrix} \in R^2$ for which $x^2 + y^2 < 1$.   On the other

hand, these points do constitute the interior of the closed

ball $\overline{N(0,1)} = \{\begin{pmatrix} x \\ y \end{pmatrix} \in R^2 \mid x^2 + y^2 \leq 1\}$.   To illustrate the idea of closure we merely note that the set $\overline{N(0,1)}$ =

$\{(\begin{smallmatrix} x \\ y \end{smallmatrix}) \in R^2 | x^2 + y^2 \le 1\}$ is the closure of $\overline{N(0,1)} = (\begin{smallmatrix} x \\ y \end{smallmatrix}) \in R^2 |$

$x^2 + y^2 < 1\}$.

We now pass from a discussion of properties of sub-

sets of R^n to a discussion of properties of <u>sequences</u> of

vectors in R^n.

DEFINITION 2.6. A <u>sequence</u> in R^n is a function whose

domain is an infinite set of non-negative integers and whose

range consists of vectors in R^n. The value of such a func-

tion at an integer k is commonly denoted by x_k, where x_k

is a vector in R^n. The domain is commonly taken to be the

set of all non-negative integers, $\{0, 1, 2, \ldots\}$, or the set of

all positive integers, $\{1, 2, 3, \ldots\}$. The range of the se-

quence is the subset of R^n: $\{x \in R^n | x = x_k$ for some k in

the domain of the sequence$\}$. We will commonly denote a

sequence by the symbol $\{x_k\}$, or if we wish to be specific

about its domain, by $\{x_k | k \in D\}$, it being understood that

D is an infinite set of non-negative integers.

DEFINITION 2.7. A sequence $\{y_k | k \in D_1\}$ is a subsequence of the sequence $\{x_k | k \in D\}$ if D_1 is an (infinite) subset of

D and $y_k = x_k$ whenever $k \in D_1$. If D_1 is a proper subset

of D then $\{y_k | k \in D_1\}$ is a proper subsequence of $\{x_k | k \in D\}$.

As an example we offer the seuqnece $\{\binom{x_k}{y_k}\}$ in R^2

defined by

$$x_k = \cos (k\frac{\pi}{3}), \qquad k = 0, 1, 2, \ldots$$

$$y_k = \sin (k\frac{\pi}{3}), \qquad k = 0, 1, 2, \ldots .$$

The domain consists of all non-negative integers while the

range consists of six points on the unit circle in R^2. The

sequence $\{\binom{x_k}{y_k}) | k = 0, 2, 4, 6, \ldots \}$ is a subsequence of the

original sequence whose range consists of three points on the

unit circle. It is entirely possible that the range of a sub-

sequence might be the same as the range of the original se-

quence. Thus $\{\binom{x_k}{y_k}) | k = 0, 5, 10, 15, \ldots \}$ is a proper sub-

sequence of $\{\binom{x_k}{y_k}), k = 0, 1, 2, 3, \ldots \}$ but in both cases the

range consists of the same six points on the unit circle.

A common way to specify a subsequence of $\{x_k\} =$

$\{x_k | k = 0, 1, 2, \ldots \}$ is to let $\{k_\ell | \ell = 0, 1, 2, \ldots \}$ be a

sequence with non-negative integer values with the property

$k_{\ell+1} > k_\ell$. The subsequence is then $\{x_k | k = k_\ell$ for some

$\ell = 0, 1, 2, \ldots$ } which is denoted by $\{x_{k_\ell}\}$. This notation swiftly gets into trouble when one speaks of subsequences of subsequences of subsequences, etc. Nevertheless, it is occasionally convenient.

DEFINITION 2.8.   A point $x \in R^n$ is the <u>limit</u> of a sequence $\{x_k | k \in D\}$ if for every $\delta > 0$ there is a positive integer k_δ such that $x_k \in N(x, \delta)$ whenever $k \in D$ and $k > k_\delta$. We write $\lim_{k \to \infty} x_k = x$.

DEFINITION 2.9.   A point $x \in R^n$ is a <u>limit point</u> of a sequence $\{x_k | k \in D\}$ if there is a subsequence $\{x_k | k \in D_1\}$ of $\{x_k | k \in D\}$ such that x is the limit of $\{x_k | k \in D_1\}$.

The sequence in R^2 defined by

$$x_k = 1 + \frac{1}{k}, \qquad k = 1, 2, 3, \ldots$$

$$y_k = -2 - \frac{1}{k^2}, \qquad k = 1, 2, 3, \ldots$$

has the point $\begin{pmatrix} 1 \\ -2 \end{pmatrix}$ as its limit. The sequence defined following Definition 2.7 has no limit but each of the points of its range is a limit point of the sequence. The sequence in R^1 defined by

$$\gamma_k = \begin{cases} 0 + \dfrac{1}{k}, & k = 1, 3, 5, 7, \dots \\[2ex] 1 - \dfrac{1}{k}, & k = 2, 4, 6, 8, \dots \end{cases}$$

has 0 and 1 as limit points but has no limit.

A sequence $\{x_k\}$ which has a limit x is said to converge to x. If we say that a sequence $\{x_k\}$ is convergent we mean that there exists a point x which is the limit of that sequence. The next theorem, though trivial to prove, establishes an important property of convergent sequences called the Cauchy Property.

THEOREM 2.2. If a sequence $\{x_k\}$ in R^n is convergent, then for every $\delta > 0$ there is a positive integer k_δ such that $\|x_k - x_\ell\| < \delta$ whenever k and ℓ are both greater than k_δ. (It being understood, of course, that k and ℓ both belong to D, the domain of the sequence. From now on this will be tacitly understood.)

Proof. Let x be the limit of $\{x_k\}$ and let k_δ be such that $x_k \in N(x, \dfrac{\delta}{2})$ whenever $k > k_\delta$. The existence of such a k_δ follows from the definition of the limit of a sequence. If k

and ℓ are both greater than k_δ we have

$$\|x_k - x_\ell\| \leq \|x_k - x\| + \|x - x_\ell\| < \frac{\delta}{2} + \frac{\delta}{2} = \delta$$

and the proof is complete.

We abbreviate the statement of the Cauchy property by writing

$$\lim_{k,\,\ell \to \infty} \|x_k - x_\ell\| = 0.$$

Thus Theorem 2.2 states that if $\lim\limits_{k \to \infty} x_k = x$ then $\lim\limits_{k,\,\ell \to \infty} \|x_k - x_\ell\| = 0$. We also express Theorem 2.2 by saying that every convergent sequence is a Cauchy sequence, a Cauchy sequence being any sequence $\{x_k\}$ for which $\lim\limits_{k,\,\ell \to \infty} \|x_k - x_\ell\| = 0$.

It is also true that every Cauchy sequence is a convergent sequence. This is not a theorem which we can prove, however. It is true that if we assume this for sequences in R^1 then we can prove it for sequences in R^n for any dimension $n > 1$. We leave this as an exercise. In R^1 the statement that every Cauchy sequence is convergent is taken as one of the defining properties of R^1, the set of all real numbers. We invoke this property when we speak of numbers represented by non-terminating decimals such as

$$\pi = 3.14159\ldots$$

$$e = 2.71828\ldots$$

$$\sqrt{2} = 1.41421\ldots\ .$$

A non-terminating decimal $.a_1a_2a_3a_4\ldots a_n\ldots$ may be
thought of as a sequence $.a_1, .a_1a_2, .a_1a_2a_3,$ etc. of terminating decimals which represent rational numbers, viz.,
$.a_1 = \dfrac{a_1}{10}, .a_1a_2 = \dfrac{a_1a_2}{100}$, etc., where the a_i are digits
0, 1, 2, 3, 4, 5, 6, 7, 8 or 9. Without difficulty we verify
that this is a Cauchy sequence. Indeed

$$\left|(.a_1a_2\ldots a_m\ldots a_n) - (.a_1a_2\ldots a_m)\right|$$

$$= .00\ldots 0a_{m+1}\ldots a_n \le .00\ldots 09\ldots 9$$

$$< 1 \times 10^{-m}.$$

We postulate the existence of a limit for this Cauchy sequence
and take it to be the real number represented by the non-terminating decimal.

Thus a sequence $\{x_k\}$ is convergent if and only if it
is a Cauchy sequence. We will use this equivalence of

convergent and Cauchy sequences many times in the sequel.

DEFINITION 2. 10. A set $S \subseteq R^n$ is <u>bounded</u> if there is a positive number B such that $\|x\| \leq B$ for all $x \in S$.

DEFINITION 2. 11. A set $S \subseteq R^n$ is <u>compact</u> if it is both closed and bounded.

DEFINITION 2. 12. Let S be a subset of R^n. A point $y \in R^n$ is a cluster point of S if every neighborhood of y includes infinitely many points in S.

We give an example in R^1. We let S be the set of all rational numbers whose absolute value is less than 1, i.e.,

$$S = \{r \in R^1 \,|\, r = \frac{p}{q}, \quad p, q \text{ integers}, \quad |p| < |q| \}.$$

We claim that every real number with absolute value ≤ 1 is a cluster point of this set. Let s be a real number with decimal expansion

$$s = . a_1 a_2 a_3 \cdots a_n \cdots .$$

For convenience we have assumed s to be non-negative.

This is clearly no real restriction. Let N be any neighbor-

hood of s in R^1. Then N must include an interval of the

form $[s - 10^{-m}, s + 10^{-m}]$ for some positive integer m.

Let r be the terminating decimal

$$r = . a_1 a_2 a_3 \cdots a_n \tilde{a}_{n+1}$$

for an integer $n \geq m$. Then

$$s - r = . 0 \ldots 0 (a_{n+1} - \tilde{a}_{n+1}) a_{n+2} \cdots \leq 10^{-n} \leq 10^{-m}$$

and hence

$$r \in [s - 10^{-m}, s + 10^{-m}] \subseteq N.$$

Now r is rational, being equal to $\dfrac{a_1 a_2 \cdots a_n \tilde{a}_{n+1}}{10^{n+1}}$, and we

have, therefore, found a rational number in the neighborhood

N of s. By varying n and the digit $\tilde{a}_{n+1}$ (the latter being

necessary only if s is a terminating decimal, i. e. if s is

rational) we can find infinitely many such r. Since N was

any neighborhood of s we have shown that s is a cluster

point of S.

Perhaps the most important theorem of this chapter is

THEOREM 2. 3. (Bolzano-Weierstrass Theorem).   Let S be a bounded subset of R^n having infinitely many elements.   Then there is at least one point $y \in R^n$ which is a cluster point of S.   If S is a closed set then $y \in S$.

Proof.   Since S is a bounded set, there is a positive number B such that $\|x\| \leq B$ for all $x \in S$.   For convenience we will assume that $B \leq 1$.   If this were not true we could prove the theorem for the set $\frac{1}{B} S = \{x \in R^n | Bx \in S\}$, and the truth of the theorem for $\frac{1}{B}S$ clearly implies that the theorem is true for S.   We may also assume S is included in $\{x \in R^n | x^k \geq 0$, $k = 1, 2, \ldots n\}$.   For if this were not true we could look instead at the set $b + S = \{x \in R^n | x - b \in S\}$, where b is the vector with components $b^k = B$, $k = 1, 2, \ldots, n$.

We consider first the case $n = 1$, assuming that the set S contains infinitely many real numbers $r = .a_1 a_2 a_3 \ldots$ $a_m \ldots$.   We divide these numbers into ten classes S_i^1, $i = 0, 1, \ldots, 9$, each S_i^1 consisting of numbers r of the given set S whose first integer is i.   Thus, e. g. , S_7^1 consists of all numbers of the form $r = .7 a_2 a_3 \ldots a_m \ldots$ which lie in S.   Now there are only ten S_i^1 and infinitely many r in S, so some $S_{i_1}^1$ must contain infinitely many such r. Now we repeat the process. $S_{i_1}^1$ can be divided into ten classes S_i^2, the members of S_i^2 being members of $S_{i_1}^1$ having the digit i in the second position. Again one of these $S_{i_2}^2$ contains infinitely many r from S.

Continuing in this way we obtain a sequence of subsets of S, $S_{i_1}^1$, $S_{i_2}^2$, $S_{i_3}^3$, ..., $S_{i_m}^m$, ..., all with infinitely many members, satisfying $S_{m+1} \subseteq S_m$, and with the property that all r in $S_{i_m}^m$ have a decimal expansion beginning with $.i_1 i_2 i_3 \ldots i_m$. We let $\hat{r}$ be the real number

$$\hat{r} = .i_1 i_2 \cdots i_m i_{m+1} \cdots$$

and note that if $r \in S_{i_m}^m$ then $|\hat{r} - r| \leq 10^{-m}$. Thus the infinite subsets $S_{i_m}^m$ of S lie within intervals $[\hat{r} - 10^{-m}, \hat{r} + 10^{-m}]$ whose length tends to 0 as m tends to ∞ and from this it is clear that every neighborhood N of $\hat{r}$ in R^1 must contain infinitely many real numbers r from S. This completes the proof for n = 1.

When n > 1 one proceeds in essentially the same way. Without loss of generality we can assume that S has infinitely many points in the set

$$P = \{x \in R^n \mid 0 \le x^i \le 1, \quad i = 1, 2, \ldots, n\}.$$

This infinite collection of points of S is divided into 10^n subclasses corresponding to all of the possibilities for initial digits in the decimal expansions of the components of x. At least one of these subclasses must still contain infinitely many points of S and we proceed to divide it into 10^n further subclasses, depending upon the second digits in the decimal expansions of the components, and so on. At the m-th stage we have an infinite subset S_m of S consisting entirely of vectors x whose components x^k all have the same digits in the first m places of their decimal expansion. Thus if x and $\hat{x}$ are in S_m their k-th components x^k and $\hat{x}^k$ both begin with $.i_1^k i_2^k \ldots i_m^k$. (Note that there is no particular relationship between different components of the same vector.) One can then easily verify that the vector

$$y = \begin{pmatrix} y^1 \\ y^2 \\ \vdots \\ y^n \end{pmatrix} = \begin{pmatrix} .i_1^1 i_2^1 \ldots i_m^1 i_{m+1}^1 \ldots \\ .i_1^2 i_2^2 \ldots i_m^2 i_{m+1}^2 \ldots \\ \vdots \\ .i_1^n i_2^n \ldots i_m^n i_{m+1}^n \ldots \end{pmatrix}$$

whose components have the indicated decimal expansions, is a cluster point of S.

If S is a closed set it includes all of its boundary points. Since every neighborhood of y includes points of S, y cannot belong to the open set S^c and must therefore belong to S. With this the proof is complete.

THEOREM 2.4. If $\{x_k\}$ is a bounded sequence in R^n, i.e., $\|x_k\| \leq B$ for all k, then $\{x_k\}$ has at least one limit point $y \in R^n$.

Proof. We consider the range $R = \{x \in R^n | x = x_k$ for some k$\}$ of the sequence $\{x_k\}$. If this set is finite, consisting of points $y_1, y_2, \ldots, y_m$, then there must be some y_i, $1 \leq i \leq m$, such that $x_k = y_i$ for infinitely many k. Then y_i is a limit point of $\{x_k\}$. If the set R is infinite then, since it is clearly bounded, it has a cluster point y by Theorem 2.3. Consider the neighborhoods $N(y, (\frac{1}{2})^j)$, $j = 1, 2, \ldots$, of y. Each such neighborhood contains infinitely many points of R. Thus $N(y, \frac{1}{2})$ contains some x_{k_1}, $N(y, (\frac{1}{2})^2)$ contains some x_{k_2}, $k_2 > k_1$, $N(y, (\frac{1}{2})^3)$ contains some x_{k_3}, $k_3 > k_2$, etc.

The resulting subsequence $\{x_{k_j}\}$ clearly has limit y and thus y is a limit point of $\{x_k\}$.

THEOREM 2.5. (Monotone convergence theorem.) Let $\{r_k\}$ be a sequence of real numbers such that $r_{k+1} \geq r_k$ for each k and $r_k \leq B$ for some B and all k. Then there is a real number $\hat{r}$ such that $\lim\limits_{k \to \infty} r_k = \hat{r}$.

<u>Proof.</u> For all k, $r_1 \leq r_k \leq B$ so $\{r_k\}$ is a bounded sequence and, by Theorem 2.4, has a limit point r. Consider the interval $N(\hat{r}, \delta) = (\hat{r} - \delta, \hat{r} + \delta)$ for $\delta > 0$. There must be a k_δ such that $r_{k_\delta} \in N(\hat{r}, \delta)$ if $\hat{r}$ is to be a limit point of r_k. Then $r_{k_\delta} \leq r_k$ for all $k > k_\delta$. But there is no $r_k \geq \hat{r} + \delta$, for if there were one, say r_{k_1}, then for all $k \geq k_1$, we would have $r_k \geq \hat{r} + \delta$ and no subsequence of $\{r_k\}$ could then converge to $\hat{r}$. Thus $\hat{r} - \delta < r_{k_\delta} \leq r_k < \hat{r} + \delta$ for all $k > k_\delta$. Since we can do this for any $\delta > 0$ we conclude that $\lim\limits_{k \to \infty} r_k = \hat{r}$ and the proof is complete.

DEFINITION 2.13. Let R be a set of real numbers. If there is a number b such that $r \leq b$ for all $r \in R$ then b is

called an <u>upper bound</u> for R. Similarly if $r \geq a$ for all $r \in R$, a is a <u>lower bound</u> for R.

DEFINITION 2.14. If $\hat{b}$ is an upper bound for R with the property that $\hat{b} \leq b$ whenever b is an upper bound for R we say that $\hat{b}$ is the <u>least upper bound</u> (l.u.b) of R. Similarly if $\hat{a}$ is a lower bound for R such that $\hat{a} \geq a$ whenever a is a lower bound for R we say that $\hat{a}$ is the <u>greatest lower bound</u> (g.l.b) of R.

For example, the set $\{r \in R^1 \mid r = \frac{1}{k}, \; k$ is a positive integer$\}$ has any non-positive number as a lower bound while 0 is the greatest lower bound.

THEOREM 2.6. If R is a non-empty set of real numbers having an upper bound b then R has a l.u.b. If R has a lower bound it has a g.l.b.

<u>Proof.</u> We will prove R has a l.u.b. if it has an upper bound. The proof that it has a g.l.b. if it has a lower bound is essentially the same. Let b be an upper bound for R and let r_1 be a point in R. Clearly $r_1 \leq b$ and if $r_1 = b$ then

b must be the l. u. b. So we assume $r_1 < b$. We then define

a sequence $\{r_k\}$ recursively by letting $d = b - r_1$ and

$$
r_{k+1} = \begin{cases} r_k + \dfrac{d}{2^k} & \text{if there is an } r \in R \text{ with } r > r_k + \dfrac{d}{2^k}, \\[2ex] r_k & \text{otherwise}. \end{cases}
$$

The sequence $\{r_k\}$ is clearly monotone non-decreasing, i. e.,

$r_{k+1} \geq r_k$, and bounded above by b. Theorem 2. 5 applies to

show that there is a real number $\hat{b}$ such that $\lim\limits_{k \to \infty} r_k = \hat{b}$

and it is clear that $r_1 \leq \hat{b} \leq b$. We claim that $\hat{b}$ is the

l. u. b. of R.

      We claim that for all k there is a number $r \in R$ such

that $r \geq r_k$ and there are no numbers $r \in R$ such that $r > r_k +$

$\dfrac{d}{2^{k-1}}$. This is clearly true for $k = 1$. Suppose it is true for

the integer $k - 1$. Thus there is some $r \in R$ such that $r \geq$

r_{k-1} and no such r with $r > r_{k-1} + \dfrac{d}{2^{k-2}}$. Consider the in-

terval $[r_{k-1}, \; r_{k-1} + \dfrac{d}{2^{k-2}}] = [r_{k-1}, \; r_{k-1} + \dfrac{d}{2^{k-1}}] \cup (r_{k-1} +$

$\dfrac{d}{2^{k-1}}, \; r_{k-1} + \dfrac{d}{2^{k-2}}] = I_k \cup J_k$. Some $r \in R$ lies in I_k or

J_k . If there is such an r in J_k then $r_k = r_{k-1} + \dfrac{d}{2^{k-1}}$ and

no $r \in R$ is such that $r > r_k + \dfrac{d}{2^{k-1}} = r_{k-1} + \dfrac{d}{2^{k-1}} + \dfrac{d}{2^{k-1}} =$

$r_{k-1} + \dfrac{d}{2^{k-2}}$. If there is no $r \in R$ in J_k then $r_k = r_{k-1}$ and no $r \in R$ is such that $r > r_{k-1} + \dfrac{d}{2^{k-1}} = r_k + \dfrac{d}{2^{k-1}}$. In either

case it is clear that there is an $r \in R$ with $r \geq r_k$.

Thus for each k, since $r_k \leq \hat{r}$, there is some $r \in R$

in $[r_k, \hat{r} + \dfrac{d}{2^{k-1}}]$ but none greater than $\hat{r} + \dfrac{d}{2^{k-1}}$. Letting

$k \to \infty$ we see that no $r \in R$ can be greater than $\hat{r}$ while

every interval $(\hat{r} - \delta, \hat{r}]$, $\delta > 0$, must contain some $r \in R$.

Then it is clear that $\hat{r}$ is the l. u. b. of R.

We are now ready to tackle the problem of the exis-

tence of maxima and minima of functions $f : R^n \to R^1$ within

sets $S \subseteq R^n$. We first remind the reader of the standard

DEFINITION 2.15. A function $f : R^n \to R^m$ is continuous at

a point $x \in R^n$ if $x \in D$ (the domain of f) and whenever

$\{x_k\}$ is a sequence of points in D with $\lim\limits_{k \to \infty} \|x - x_k\| = 0$

we also have $\lim\limits_{k \to \infty} \|f(x) - f(x_k)\| = 0$. ($\|x - x_k\|$ refers to

the norm in R^n, $\|f(x) - f(x_k)\|$ to the norm in R^m.)

For real valued functions $f : R^n \to R^1$ we can define

another type of continuity which is occasionally useful.

DEFINITION 2.16. A function $f : R^n \to R^1$, defined on a do-

main $D \subseteq R^n$, is said to be <u>lower semi-continuous</u> at a point

$x \in D$ if whenever $\{x_k\}$ is a sequence of points in D with $\lim_{k \to \infty} \|x - x_k\| = 0$ we have

$$\liminf_{k \to \infty} f(x_k) \geq f(x)$$

where $\liminf_{k \to \infty} f(x_k)$ is defined as the g. l. b. of the set of limit points of $\{f(x_k)\}$. In the same way we say that f is upper semi-continuous at x if

$$\limsup_{k \to \infty} f(x_k) \leq f(x)$$

where $\limsup_{k \to \infty} f(x_k)$ is defined as the l. u. b. of the set of limit points of $\{f(x_k)\}$.

It is not hard to show that if f is both upper and lower semi-continuous at x then f is continuous at x. If f is continuous at all points of D we say f is continuous in D and we apply the same convention to upper and lower semi-continuity.

A simple example of a lower semi-continuous function is

$$f(x) = \begin{cases} 1, & x \neq 0 \\ 0, & x = 0 \end{cases}$$

which is lower semi-continuous at 0. The function $-f(x)$ is upper semi-continuous at 0. Indeed, $-f(x)$ is always upper semi-continuous if $f(x)$ is lower semi-continuous and vice versa.

THEOREM 2.7. If $f : R^n \to R^1$ is lower semi-continuous on a compact set $S \subseteq R^n$ then there is a real number B such that $f(x) \geq B$ for all $x \in S$.

Proof. Suppose this were not true. Then for every integer $k > 0$ there would be at least one point $x_k \in S$ with $f(x_k) \leq -k$. Now $\{x_k\}$, since it is confined to the compact, i.e., closed and bounded, set S has at least one limit point $y \in R^n$ by Theorem 2.4. Since S is closed, $y \in S$. Let $\{x_{k_\ell}\}$ be a subsequence of $\{x_k\}$ with limit y. Then $f(y)$ is defined as some finite real number and we must have, by lower semi-continuity

$$\liminf_{\ell \to \infty} f(x_{k_\ell}) \geq f(y).$$

But this is impossible because $\lim_{\ell \to \infty} f(x_{k_\ell}) \leq \lim_{\ell \to \infty} (-k_\ell) = -\infty$.

Thus we have a contradiction and the theorem must be true.

COROLLARY. If $f : R^n \to R^1$ is upper semi-continuous on a compact set S there exists a real number B such that $f(x) \le B$ for all $x \in S$. If f is continuous on a compact set S there is a real number $B > 0$ such that $|f(x)| \le B$ for all $x \in S$.

Proof. If f is upper semi-continuous, -f is lower semi-continuous and thus $-f(x) \ge -B$ for some real number $- B$. Then for all $x \in S$, $f(x) \le B$. If f is continuous it is both upper and lower continuous so that both f and -f are bounded above by some $B > 0$, whence $|f(x)| \le B$ for $x \in S$.

THEOREM 2.8. If f is lower semi-continuous on a compact set $S \subseteq R^n$ there is a point $x^* \in S$ such that $f(x) \ge f(x^*)$ for all $x \in S$ so that x^* is a minimum of f in S.

Proof. Theorem 2.6 together with Theorem 2.7 show that the set $\{r \in R^1 | r = f(x)$ for some $x \in S\}$ has a g. l. b. $\hat{b}$. We claim there is a point $x^* \in S$ such that $f(x^*) = \hat{b}$. If this were not true then $g(x) = f(x) - \hat{b}$ would be a lower semi-continuous function assuming only positive values in S. Then it is easy to see that $\frac{1}{g(x)}$ is upper semi-continuous in S

and, by Theorem 2.7, we must have $\frac{1}{g(x)} \leq B$ for some $B > 0$.

But then $g(x) = f(x) - \hat{b} \geq \frac{1}{B}$ for all $x \in S$ so that $f(x) \geq \hat{b} +$

$\frac{1}{B} > \hat{b}$ for all $x \in S$. But this contradicts our choice of $\hat{b}$ as

the g. l. b. of the set described above. Thus there must be a

point $x^* \in S$ where $f(x^*) = \hat{b}$ and such a point is clearly a

(global) minimum of f in S.

COROLLARY. If f is upper semi-continuous on a compact

set $S \subseteq R^n$ then there is a point $x^* \in S$ such that $f(x) \leq$

$f(x^*)$ for all $x \in S$ so that x^* is a maximum for f in S.

If f is continuous on a compact set S then there are points

x^* and $\hat{x}$ in S with $f(\hat{x}) \leq f(x) \leq f(x^*)$ for all $x \in S$ so

that f achieves both a maximum and a minimum value in S.

Thus, at long last, we provide an answer for question

(a) of Chapter 1. There are other conditions which one could

impose upon f and S to guarantee the existence of minima

or maxima but they would either be of a rather special nature

or else would essentially reduce to those we have used.

Theorem 2.8 and its Corollary represent <u>sufficient</u>

conditions for the existence of minima and maxima. They are

not <u>necessary</u> conditions, i. e. , f may achieve minima or

maxima in a set S even if none of the conditions of Theorem

2. 8 and its Corollary are satisfied.   Thus the function   ·

$$f(r) = \begin{cases} \dfrac{1}{r} + r, & r \neq 0 \\[2em] 3, & r = 0 \end{cases}$$

has a global minimum in the set $S = \{r \in R^1 | r > -\frac{1}{3}\}$ at the

point $r^* = 1$ even though f is neither bounded nor contin-

uous on S and S is neither closed nor bounded.

EXERCISES.   CHAPTER 2

1.   Show that if every Cauchy sequence in R^1 has a limit

then every Cauchy sequence in R^n, $n > 1$, must also

have a limit.   This property of R^n is expressed by say-

ing that R^n is <u>complete</u>.

2.   Assume that $f : R^n \rightarrow R^1$ is continuous on R^n.   Show that

the set $\{x \in R^n | f(x) < c\}$, where c is a fixed real num-

ber, is an open set whereas the sets $\{x \in R^n | f(x) \leq c\}$

and $\{x \in R^n | f(x) = c\}$ are closed.

3.   Let S_1, S_2, S_3, ... be an infinite collection of subsets

of R^n. We define the union of all of these sets by

$$\bigcup_{k=1}^{\infty} S_k = \{x \in R^n | x \in S_k \text{ for some } k\}.$$

Show that if all of the S_k are open sets then $\bigcup_{k=1}^{\infty} S_k$ is

open. What can you say if all of the S_k are closed?

Consider, e.g., $S_k = [0, 1 - \frac{1}{k}]$, $k = 1, 2, 3, ...$ in R^1.

How would you define $\bigcap_{k=1}^{\infty} S_k$ and what analogous pro-

perties does it possess?

4.   Identify the cluster points (if any) of the following sets.

(i) $\{r \in R^1 | r = (-1)^k (1 - \frac{1}{k})$, for some positive integer $k\}$

(ii) $\{r \in R^1 | r$ is an integer$\}$

(iii) $\{r \in R^1 | r = \frac{j}{2^k}$ for some positive integers j and $k\}$

(iv) $\{r \in R^1 | r = r_k$ for some k, where $r_1 = 1$ and the

r_k are defined recursively by $r_{k+1} = r_k - (\frac{r_k^2 - 2}{2r_k})$,

$k \geq 1\}$

(v) $\{(\begin{smallmatrix} r \\ \theta \end{smallmatrix}) \in R^2$ (polar coordinates)$| r = r_k$, $\theta = \theta_k$ for

some k, $r_1 = 1$, $\theta_1 = 0$, $r_{k+1} = r_k + (\frac{1}{2})^k$, $\theta_{k+1} = $

$\theta_k + \frac{2\pi}{k}\}$.

5.   Consider the problem: find the square of largest area

which is contained in the regular hexagon whose sides

have length 1. (See Figure 2.)

(i)   Does this problem have a solution and is it unique?

(ii)   Find this largest area.

(iii) Replace "hexagon" by heptagon and do either (i) or

       (ii), whichever you believe to be easier.

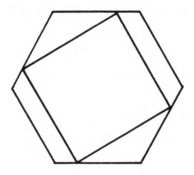

Figure 2

CHAPTER 3. ONE DIMENSIONAL BLOCK
SEARCH TECHNIQUES

In this chapter we will consider what are called <u>search</u> techniques for finding minima or maxima of <u>unimodal</u> functions f defined on a finite interval $[a, b] \subseteq R^1$. By a search technique we mean a program which reduces the possible locations of minima or maxima of f in $[a,b]$ to smaller and smaller intervals by means of evaluation of $f(r)$ at judiciously chosen points r in $[a,b]$.

DEFINITION 3.1. A function $f : R^1 \to R^1$ is $(+)$ unimodal on an interval $[a,b]$ if there is a point $r^* \in [a,b]$ such that (i) if r_1 and r_2 belong to $[a,r^*]$ and $r_1 < r_2$ then $f(r_1) < f(r_2)$; (ii) if r_1 and r_2 belong to $[r^*, b]$ and $r_1 < r_2$ then $f(r_1) > f(r_2)$. The function f is called $(-)$

unimodal if $-f$ is $(+)$ unimodal.

If f is $(+)$ unimodal on $[a, b]$ it is clear that r^* is the unique maximum of f in $[a, b]$. If f is $(-)$ unimodal then r^* is the unique minimum of f in $[a, b]$. We do not rule out in Definition 3.1 the possibility that $r^* = a$ or $r^* = b$. In these cases f is either monotone decreasing or monotone increasing. Note that no continuity assumptions are made on f. To provide an example we note that $f(\theta) = \sin \theta$ is $(+)$ unimodal on $[-\frac{\pi}{2}, \frac{3\pi}{2}]$, $(-)$ unimodal on $[\frac{\pi}{2}, \frac{5\pi}{2}]$ but neither $(+)$ nor $(-)$ unimodal on $[0, 2\pi]$.

In this chapter we will discuss procedures for finding the maximum $r^* \epsilon [a, b]$ of a function f which is $(+)$ unimodal on $[a, b]$. The analogous procedures for finding the minimum of a $(-)$ unimodal function will be quite evident. By an $\underline{\text{n-block}}$ we will mean a set of n distinct points $r_1, r_2, \ldots,$ r_n in the interior (a, b) of $[a, b]$ with $r_k < r_{k+1}$, $k = 1, 2, \ldots,$ $n-1$. Our search techniques are based on

THEOREM 3.1.   Let f be $(+)$ unimodal on $[a, b]$ and let $\{r_1, r_2, \ldots, r_n\}$ be an n-block in (a, b). Agree to denote a by r_0 and b by r_{n+1}. Among the finitely many numbers

$f(r_k)$, $k = 1, 2, \ldots, n$ let $f(r_{k_0})$ be the largest, i. e. ,

$f(r_k) \leq f(r_{k_0})$, $k = 1, 2, \ldots, n$. Then

(i) if $f(r_k) < f(r_{k_0})$, $k \neq k_0$, $r^* \epsilon [r_{k_0 - 1}, r_{k_0 + 1}]$;

(ii) if there is a $k_1 \neq k_0$ such that $f(r_{k_1}) = f(r_{k_0})$ then $k_1 = k_0 + 1$ or $k_0 - 1$, in which case r^* belongs to $[r_{k_0}, r_{k_0 + 1}]$ or $[r_{k_0 - 1}, r_{k_0}]$, respectively.

Proof. Assume (i) holds. If $r^* \notin [r_{k_0 - 1}, r_{k_0 + 1}]$ then either $r^* < r_{k_0 - 1}$ or $r^* > r_{k_0 + 1}$. Suppose the first of these to be true. Then $r_{k_0 - 1}$ and r_{k_0} both belong to $[r^*, b]$ and $r_{k_0 - 1} < r_{k_0}$. But $f(r_{k_0 - 1}) < f(r_{k_0})$, contradicting (ii) in Definition 3.1. Thus $r^* < r_{k_0 - 1}$ cannot hold. In essentially the same way we show $r^* > r_{k_0 + 1}$ cannot hold so $r^* \epsilon [r_{k_0 - 1}, r_{k_0 + 1}]$.

If (ii) holds we argue as follows. Suppose $k_1 < k_0 - 1$. Then $f(r_{k_0 - 1}) \leq f(r_{k_0}) = f(r_{k_1})$ and clearly $r^* \neq r_{k_0 - 1}$ so either $r^* < r_{k_0 - 1}$ or $r^* > r_{k_0 - 1}$. If $r^* < r_{k_0 - 1}$ then $r_{k_0 - 1}$ and r_{k_0} lie in $[r^*, b]$, $r_{k_0 - 1} < r_{k_0}$ but $f(r_{k_0 - 1}) \leq f(r_{k_0})$, contradicting (ii) of Definition 3.1. If $r^* > r_{k_0 - 1}$ then r_{k_1} and $r_{k_0 - 1}$ both lie in $[a, r^*]$, $r_{k_1} < r_{k_0 - 1}$ but $f(r_{k_1}) \geq f(r_{k_0 - 1})$, contradicting (i) of Definition 3.1. Thus we cannot have

$k_1 < k_0 - 1$. Similarly we cannot have $k_0 > k_0 + 1$ so $k_1 =$

$k_0 - 1$ or $k_1 = k_0 + 1$. We may assume without loss of gen-

erality that $r_{k_1} = r_{k_0-1}$ since the roles of r_{k_1} and r_{k_0} are

interchangeable. Now if r^* does not belong to $[r_{k_0-1}, r_{k_0}]$

either $[a, r^*]$ or $[r^*, b]$ must contain both r_{k_0-1} and

r_{k_0} where f has equal values thus contradicting either (i)

or (ii) of Definition 3.1. We conclude that $r^* \epsilon [r_{k_0-1}, r_{k_0}]$

and the proof is complete.

Our search technique consists of the following steps:

(i)   evaluate f at points $r_1 < r_2 < \ldots < r_{n-1} < r_n$

      in (a, b);

(ii) find r_{k_0} such that $f(r_k) \leq f(r_{k_0})$, $k = 1, 2, \ldots, n$;

(iii) if $f(r_k) < f(r_{k_0})$ for $k \neq k_0$, let $a_1 = r_{k_0-1}$,

      $b_1 = r_{k_0+1}$;

(iv) if $f(r_{k_0-1}) = f(r_{k_0})$, let $a_1 = r_{k_0-1}$, $b_1 = r_{k_0}$;

(v)   replace the original interval $[a, b]$ by $[a_1, b_1]$,

      form a further n-block in (a_1, b_1) and repeat the

      above steps.

In this way we obtain successively smaller intervals $[a_k, b_k]$

such that $r^* \epsilon [a_k, b_k]$, $k = 1, 2, 3, \ldots$. Moreover, we have the inclusion property $[a_{k+1}, b_{k+1}] \subseteq [a_k, b_k]$, k =

0, 1, 2, 3, ... , agreeing to let $a = a_0$, $b = b_0$. Thus the

sequences $\{a_k\}$ and $\{b_k\}$ are non-decreasing and non-

increasing, respectively. It is very easy to give conditions

under which both of these sequences coverge to r^*.

THEOREM 3. 2.   Let the successive n-blocks $\{r_{k,1}, r_{k,2}, \cdots$

$r_{k,n}\}$, k = 0, 1, 2, 3, ... be chosen in such a way that

$$r_{k,j+1} - r_{k,j-1} \leq L(b_k - a_k), \quad j = 1, 2, \dots. n, \; k = 0, 1, 2, \dots,$$

where L is a constant, $0 < L < 1$.   Then for all k

$$b_k - a_k \leq L^k(b - a)$$

so that $\lim\limits_{k \to \infty} a_k = \lim\limits_{k \to \infty} b_k = r^*$. If after k   such steps we

take $\dfrac{a_k + b_k}{2}$ as our approximation to r^* then

$$\left| r^* - \frac{a_k + b_k}{2} \right| \leq \frac{L^k}{2} (b - a).$$

Proof.   Since $[a_{k+1}, b_{k+1}]$ is one of the intervals $[r_{k,j-1},$

$r_{k,j+1}]$, j = 1, 2, ... , n, or else one of the intervals $[r_{k,j-1},$

$r_{k,j}]$, j = 2, 3, ... , n the hypotheses of the theorem guarantee that

$$b_{k+1} - a_{k+1} \le L(b_k - a_k).$$

Since this is true for all $k = 0, 1, 2, \ldots$

$$(b_k - a_k) \le L(b_{k-1} - a_{k-1}) \le L^2(b_{k-2} - a_{k-2}) \le \ldots \le L^k(b_0 - a_0)$$

and thus $(b_k - a_k) \le L^k(b - a)$. The sequences $\{a_k\}$ and $\{b_k\}$ being non-decreasing and non-increasing, respectively have limits r_a and r_b. Since for $k = 0, 1, 2, \ldots$

$$a_k \le r_a \le r_b \le b_k$$

we must have $r_b - r_a \le L^k(b - a)$. Letting $k \to \infty$, the fact that $0 < L < 1$ implies $r_b = r_a$ so that both sequences have the same limit $\hat{r}$. For each k both r^* and $\hat{r}$ lie in $[a_k, b_k]$ so $|r^* - \hat{r}| \le L^k(b - a)$ for all k and thus $\hat{r} = r^*$. Thus $\lim_{k \to \infty} a_k = \lim_{k \to \infty} b_k = r^*$. Finally, for each k, r^* lies in

$$[a_k, b_k] = [a_k, \frac{a_k + b_k}{2}] \cup [\frac{a_k + b_k}{2}, b]$$ and thus in one of the latter two intervals. The length of either of these intervals is $\frac{1}{2}(b_k - a_k) \le \frac{L^k}{2}(b - a)$ so $|r^* - \frac{a_k + b_k}{2}| \le \frac{L^k}{2}(b - a)$ and with this the proof of our theorem is complete.

EXAMPLE. The patent medicine hawker at the county fair

knows that if he charges $1 per bottle one thousand people

will buy his Old Blackfoot Remedy.   Experience tells him that

if he doubles his price he will cut that number to $250 = \dfrac{1000}{2^2}$,

if he triples the price $\dfrac{1000}{3^2}$ = 111 will buy, etc.   Assuming

the bottles cost him 80¢ each, what price should be charge

to realize maximum profit ?

     At $1 per bottle our entrepreneur makes $200.   At $3

per bottle he makes ($2.20) × (111) = $244.20 and it is

fairly clear that the take will be downhill thereafter.   For a

price $r, $1 \le r \le 3$, the profit is, in dollars,

$$p(r) = (r - \frac{4}{5})\frac{1000}{r^2} = \frac{1000}{r} - \frac{800}{r^2}.$$

As our hawker knows no calculus, he elects to use a search

technique rather than differentiation.   The search technique

will consist of 3-blocks with the points evenly spaced in

each interval.

Step 1.

	$r_1 = 3/2$	$r_2 = 2$	$r_3 = 5/2$
$p(r) =$	$\dfrac{1000}{(3/2)} - \dfrac{800}{(9/4)}$	$\dfrac{1000}{2} - \dfrac{800}{4}$	$\dfrac{1000}{(5/2)} - \dfrac{800}{(25/4)}$
	$= 311$	$= 300$	$= 272$

(All figures rounded to nearest dollar.)

Since the maximum occurs at $r_1 = 3/2$ and nowhere else the new interval is $[a_1, b_1] = [1, 2]$.

Step 2

$r_1 = 5/4$	$r_2 = 3/2$	$r_3 = 7/4$
$p(r) = \dfrac{1000}{(5/4)} - \dfrac{800}{(25/16)}$	Computed in Step 1	$\dfrac{1000}{(7/4)} - \dfrac{800}{(49/16)}$
$= 288$	$= 311$	$= 311$

Since the profits at $r_2 = 3/2$ and $r_3 = 7/4$ are equal, (ii) of Theorem 3.1 is applied to get $[a_2, b_2] = [3/2, 7/4]$.

Step 3

$r_1 = 25/16$	$r_2 = 13/8$	$r_3 = 27/16$
$p(r) = \dfrac{1000}{\left(\frac{25}{16}\right)} - \dfrac{800}{\left(\frac{625}{256}\right)}$	$\dfrac{1000}{\left(\frac{13}{8}\right)} - \dfrac{800}{\left(\frac{169}{64}\right)}$	$\dfrac{1000}{\left(\frac{27}{16}\right)} - \dfrac{800}{\left(\frac{729}{256}\right)}$
$= 312$	$= 313$	$= 312$

The new interval is $[a_3, b_3] = [\frac{25}{16}, \frac{27}{16}]$.

Step 4

$r_1 = 51/32$	$r_2 = 13/8$	$r_3 = 53/32$
$p(r) = \dfrac{1000}{\left(\frac{51}{32}\right)} - \dfrac{800}{\left(\frac{2601}{1024}\right)}$	Computed in Step 1	$\dfrac{1000}{\left(\frac{52}{32}\right)} - \dfrac{800}{\left(\frac{2809}{1024}\right)}$
$= 312$	$= 313$	$= 312$

At this point he terminates his computations as it seems highly unlikely that he can do better than $ 313. The optimal selling price has been found to be $\$\frac{13}{8} = \$1.63$.

The reader should note that the computation in both Step 2 and Step 4 was simplified because the midpoint of the new interval was in each case one of the points of the preceding 3-block. Note also that the choice of $[a_2, b_2]$ was made at some hazard. If we had rounded to cents rather than dollars the costs at $\frac{3}{2}$ and $\frac{7}{4}$ in Step 2 would not have been exactly equal and we would have chosen either $[\frac{5}{4}, \frac{7}{4}]$ or $[\frac{3}{2}, 2]$ as $[a_2, b_2]$.

We now take up the subject of <u>optimal</u> search policies. Given that we are going to use a sequence of n-blocks in order to confine the maximum of a (+) unimodal function f to smaller and smaller intervals, the rate at which they become small being governed by the relationships outlined in Theorem 3.2, how should we place the points $r_1, r_2, \ldots, r_n$ in each block? That is, what arrangement will make the constant L of Theorem 3.2 as small as possible? Perhaps surprisingly, the answer depends very strongly upon whether n, the number of points, is even or odd. When we have obtained this result we can pass to the question of how we should choose n for maximum efficiency. Here again the answer is not entirely obvious.

THEOREM 3. 3. If $n = 2m - 1$ is odd the minimum value of L is $\frac{1}{m} = \frac{2}{n+1}$ and this value of L can be attained by spacing the points $r_1, r_2, \ldots, r_n$ evenly in $[a, b]$, i. e.

$$r_k = a + \frac{k(b-a)}{n+1}, \quad k = 1, 2, \ldots, n.$$

This spacing is not the only one yielding this value of L; there are infinitely many. If $n = 2m$ is even then $L > \frac{1}{m+1} = \frac{2}{n+2}$ for every arrangement of the points $r_1, r_2, \ldots, r_n$. The value $L = \frac{2}{n+2}$ cannot be achieved but we may approach this value as closely as we wish by spacing points $\rho_1, \rho_2, \ldots, \rho_m$ evenly in $[a, b]$, i. e.

$$\rho_k = a + \frac{k(b-a)}{m+1}, \quad k = 1, 2, \ldots, m$$

and taking $r_{2k-1} = \rho_k - \delta$,

$$r_{2k} = \rho_k + \delta, \quad j = 1, 2, \ldots, m,$$

for some small $\delta > 0$. This arrangement yields the value

$$L = \frac{2}{n+2} + \frac{\delta}{b-a}.$$

<u>Proof.</u> Assume $n = 2m - 1$ is odd. Since $a < r_1 < r_2 < \ldots <$ $r_{2m-2} < r_{2m-1} < b$ we have

$$(b - r_{2m-2}) + (r_{2m-2} - r_{2m-4}) + \ldots + (r_4 - r_2) + (r_2 - a) = b - a$$

and each of the terms in the sum is positive. Now whenever

we have m positive numbers $p_1, p_2, \ldots, p_m$ with sum q it

is clear that at least one of the $p_i \geq \frac{q}{m}$ and if any $p_j < \frac{q}{m}$

some $p_i > \frac{q}{m}$. Returning to the case at hand we see that at

least one of the numbers $r_{2k} - r_{2k-2} \geq \frac{b-a}{m}$ and strict in-

equality will hold for some k unless all numbers $r_{2k} - r_{2k-2}$,

$k = 1, 2, \ldots, m$, are equal to $\frac{b-a}{m}$, i. e. unless $r_{2k} = a +$

$\frac{2k(b-a)}{n+1}$, $k = 1, 2, \ldots, m$. Thus it is clear that L cannot be

less than $\frac{1}{m} = \frac{2}{n+1}$ and that we cannot even have $L = \frac{1}{m}$

unless the even indexed r_i are evenly spaced. Assuming

now that the even indexed r_k are evenly spaced, choose r_1

to be any point such that $a < r_1 < r_2$ and let $r_{2k+1} = r_{2k-1} +$

$\frac{(b-a)}{m}$, $k = 1, 2, \ldots, m-1$. It is clear then that each interval

$r_{k+1} - r_{k-1}$, $k = 1, 2, \ldots, n$, has length exactly equal to $\frac{b-a}{m}$

so that $L = \frac{1}{m}$ for this arrangement of the r_k. If we take

$r_1 = a + \frac{b-a}{2m}$ the r_k will all be evenly spaced. This arrange-

ment is clearly most convenient and has certain advantages

which we will point out soon. The arrangement for an arbitrary choice of r_1 in (a, r_2) is shown in Figure 3.

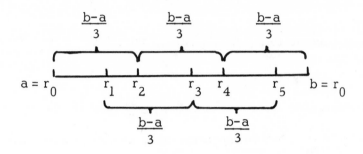

Figure 3

Now suppose $n = 2m$ is even. Let the r_k be arranged so that $a < r_1 < r_2 < \ldots < r_{2m-1} < r_{2m} < b$ and let

$$\rho_k = \frac{r_{2k} + r_{2k-1}}{2}, \quad k = 1, 2, \ldots, m. \quad \text{Then}$$

$$(b - \rho_m) + (\rho_m - \rho_{m-1}) + \ldots + (\rho_2 - \rho_1) + (\rho_1 - a) = b - a$$

and thus one of the terms on the left hand side must be greater than or equal to $\frac{b-a}{m+1} = \frac{2(b-a)}{n+2}$. We let $\delta_k = \frac{r_{2k} - r_{2k-1}}{2}$ and examine the intervals $r_{k+1} - r_{k-1}$, $k = 1, 2, \ldots, n$. Suppose $b - \rho_m \geq \frac{b-a}{m+1}$. Then, since $r_{2m-1} < \rho_m$ we have

$$r_{2m+1} - r_{2m-1} = b - r_{2m-1} > \frac{b-a}{m+1}.$$ A similar situation occurs

if $r_2 - a \geq \frac{b-a}{m+1}$. On the other hand, suppose some

$\rho_k - \rho_{k-1} \geq \frac{b-a}{m+1}$, $k = 2, 3, \ldots, m$. We note that

$$(r_{2k} - r_{2k-2}) + (r_{2k-1} - r_{2k-3}) = (\rho_k + \delta_k - (\rho_{k-1} + \delta_{k-1}))$$

$$+ (\rho_k - \delta_k - (\rho_{k-1} - \delta_{k-1})) = 2(\rho_k - \rho_{k-1}) \geq \frac{2(b-a)}{m+1}$$

which implies that one of the numbers $r_{2k} - r_{2k-2}$, $r_{2k-1} - r_{2k-3}$ is greater than or equal to $\frac{b-a}{m+1}$ and equality can hold only if both equal $\frac{b-a}{m+1}$. If one of these is greater than $\frac{b-a}{m+1}$, $k = 2, 3, \ldots, m$, it is clear that we cannot achieve $L = \frac{1}{m+1}$. On the other hand, if $r_{2k} - r_{2k-2} = \frac{b-a}{m+1}$, $r_{2k-1} - r_{2k-3} = \frac{b-a}{m+1}$ for $k = 2, 3, \ldots, m$ we note that all of the numbers δ_k must be equal to a constant $\delta > 0$ and $\rho_k - \rho_{k-1}$ must equal $\frac{b-a}{m+1}$ for $k = 2, 3, \ldots, m$. But then, since

$$(b - \rho_m) + (\rho_m - \rho_{m-1}) + \ldots + (\rho_2 - \rho_1) + (\rho_1 - a) = b - a \quad \text{we}$$

must have

$$(b - \rho_m) + (\rho_1 - a) = b - a - (m-1)(\frac{b-a}{m+1}) = \frac{2(b-a)}{m+1}$$

so that either $b - \rho_m \geq \frac{b-a}{m+1}$ or $\rho_1 - a \geq \frac{b-a}{m+1}$. Then either $b - r_{2m-1} > \frac{b-a}{m+1}$ or $r_2 - a > \frac{b-a}{m+1}$ showing finally that it is impossible to achieve $L = \frac{1}{m+1} = \frac{2}{n+2}$. Spacing the ρ_k

evenly and putting $r_{2k-1} = \rho_k - \delta$, $r_{2k} = \rho_k + \delta$, the inter-

vals $[a, r_2]$, $[r_1, r_3]$, $[r_2, r_4]$, ..., $[r_{2m-2}, r_{2m}]$, $[b - r_{2m-1}]$

are easily seen to have lengths $\dfrac{2(b-a)}{n+2} + \delta$, $\dfrac{2(b-a)}{n+2}$, $\dfrac{2(b-a)}{n+2}$,

..., $\dfrac{2(b-a)}{n+2}$, $\dfrac{2(b-a)}{n+2} + \delta$ so that $L = \dfrac{2}{n+2} + \dfrac{\delta}{b-a}$. This

arrangement is shown in Figure 4. With this the proof of the

theorem is complete.

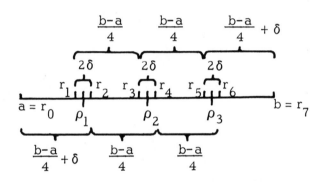

Figure 4

A remark is in order concerning the above theorem. If

we have some idea in advance where r^* lies in $[a, b]$ it

would not be optimal to space the points $r_1, r_2, ..., r_n$ as

indicated by the theorem. The thing to do would be to con-

centrate the r_k in the region where we expect r^* to lie.

The placement of the r_k indicated in Theorem 3.3 is based

on the assumption that we have no idea whatever where r^*

lies and thus we wish to make the <u>largest</u> of the numbers

$r_{k+1} - r_{k-1}$ as small as possible. We have also neglected

Case (ii) of Theorem 3. 1 since it will essentially never occur

though it is a possibility.

We are now ready to combine the results of Theorems

3. 2 and 3. 3 to obtain an estimate of the number of blocks, i.e.,

steps, necessary to find r^* within a given tolerance ϵ.

Noting the remark at the end of Theorem 3. 2 we see that k

steps are sufficient to guarantee that our error is $\leq \epsilon$ if

$b_k - a_k \leq 2\epsilon$.   Then since $b_k - a_k \leq L^k (b-a)$ all we need is

$$L^k (b - a) \leq 2\epsilon .$$

Taking logarithms (to base 10) we require

$$k \log L + \log (b-a) \leq \log \epsilon + \log 2 = \log \epsilon + .\,30103 .$$

Since $0 < L < 1$ we have $\log L < 0$ so we require

$$k \geq \frac{\log (b-a) - \log \epsilon - .\,30103}{- \log L} .$$

Thus the required number of steps is the least positive integer

k for which this inequality holds. If n is odd the formula

to use, assuming we place the points $r_1, r_2, \ldots, r_n$ optimally, is

$$k \geq \frac{\log (b-a) - \log \epsilon - .30103}{\log (n+1) - .30103} .$$

If k is even we use

$$k \geq \frac{\log (b-a) - \log \epsilon - .30103}{- \log (\frac{2}{n+2} + \frac{\delta}{b-a})} .$$

If $\frac{\delta}{b-a}$ is quite small compared with $\frac{2}{n+2}$ it will ordinarily be sufficient to take

$$k \geq \frac{\log (b-a) - \log \epsilon - .30103}{\log (n+2) - .30103} .$$

EXAMPLE. We return to the patent medicine man of our earlier example. Assuming that the profits realized from selling at various prices r are computed exactly, how many steps should it take, using 3-block search, to find the optimal selling price within $\frac{1}{2}$ cent? In this case $n = 3$, $b - a = 3 - 1 = 2$ and $\epsilon = .005$. Thus

$$k \geq \frac{\log 2 - \log (.005) - \log 2}{\log 4 - \log 2} = \frac{3 - \log 5}{\log 2}$$

$$= \frac{3 - .699}{.301} = \frac{2.30}{.301} = 7.65 .$$

and we conclude that 8 steps would be required for this degree of accuracy.

In actual fact it would be quite laborious to carry out these 8 steps by hand. In order to properly discriminate between the profits in the later steps we would need to compute these quite accurately. We will have more to say about this difficulty soon.

We now tackle the question of the relative efficiency of search techniques using different values of n, i. e. , employing different numbers of points per block. When we do this we must have a clear idea of what efficiency is to be judged by. The most natural criterion would seem to be the number of times we have to evaluate the function f in order to obtain r^* within a given tolerance ϵ. Before computing this number we must make some conventions.

We will assume that the n points in an n-block are optimally placed and that if n is odd the points are evenly spaced. This latter point is quite important. As we have indicated, we must assume that it is case (i) of Theorem 3.1 which always applies. If we always use evenly spaced points in the case where n is odd the point r_{k_0} where f is maximal at the j-th step will be one of the points of the n-block used

at the (j+1)-st step. This means one of the function evalua-

tions of the (j+1)-st step has already been performed at the

j-th step, resulting in a saving of one function evaluation.

This can be observed at Steps 2 and 4 of the example follow-

ing Theorem 3.2. Thus, for n odd, n function evaluations

are required at the first step but only n-1 at each succeed-

ing step.

When n is even we have the number δ to worry a-

bout. We will assume that δ is negligibly small relative to

$\frac{2}{n+2}$ and will neglect it in our present computations.

We see then that to achieve an accuracy of ϵ, odd-

block (n odd) search techniques require $n + (n-1)(k-1)$ func-

tion evaluations whereas even-block (n even) search tech-

niques require nk such evaluations. If we assume that ϵ is

very small compared with $\frac{1}{n}$ we can adequately represent the

number of function evaluations, N, by

$$N = n + (n-1) \left(\frac{\log\left(\frac{b-a}{2}\right) - \log \epsilon}{\log\left(\frac{n+1}{2}\right)} \right) \quad , \quad n \text{ odd}$$

$$N = n \left(\frac{\log\left(\frac{b-a}{2}\right) - \log \epsilon}{\log\left(\frac{n+2}{2}\right)} \right) \quad , \quad n \text{ even.}$$

Since $\log(\frac{b-a}{2}) - \log \epsilon$ does not depend upon n, the efficiency of the procedure can be judged from the quantities

$$\hat{N} = \frac{n-1}{\log(\frac{n+1}{2})} \quad , \quad n \text{ odd},$$

(We have assumed ϵ very small so that $\frac{n}{\log(b-a) - \log \epsilon}$ is negligible.)

$$\hat{N} = \frac{n}{\log(\frac{n+2}{2})} \quad , \quad n \text{ even}.$$

Let us first of all compare adjacent even $(n = 2m)$ and odd $(n = 2m+1)$ values of n. We have

$$\hat{N}(2m+1) = \frac{(2m+1)-1}{\log(\frac{(2m+1)+1}{2})} = \frac{2m}{\log(\frac{2m+2}{2})}$$

$$\hat{N}(2m) = \frac{2m}{\log(\frac{2m+2}{2})} \quad .$$

Thus we see that if ϵ is very small compared with $\frac{1}{n}$, odd-block search with $n = 2m+1$ is just about as efficient as even block search with $n = 2m$. In fact, if δ is negligible, exactly one more evaluation is required with $n = 2m+1$ as compared with $n = 2m$.

To decide whether n should be large or small we investigate the function $\hat{N}(\mu) = \dfrac{\mu}{\log(\frac{\mu+2}{2})}$. Differentiating we have

$$\hat{N}'(\mu) = \frac{\log_{10}(\frac{\mu+2}{2}) - \mu\frac{d}{d\mu}(\log_{10}(\frac{\mu+2}{2}))}{(\log_{10}(\frac{\mu+2}{2}))^2}$$

$$= \frac{\log_{10}(\frac{\mu+2}{2}) - \mu(\log_{10}e)\frac{d}{d\mu}(\log_e(\frac{\mu+2}{2}))}{(\log_{10}(\frac{\mu+2}{2}))^2}$$

$$= \frac{\log_{10}(\frac{\mu+2}{2}) - \dfrac{\mu \log_{10}e}{\mu+2}}{(\log_{10}(\frac{\mu+2}{2}))^2}$$

$$= \frac{\log_{10}\alpha - (1-\frac{1}{\alpha})\log_{10}e}{(\log_{10}(\alpha))^2} \, , \quad \text{where } \alpha = \frac{\mu+2}{2}$$

which, as indicated in Figure 5, is always positive for $\alpha \geq 2$. Thus in the interval $\mu \geq 2$ the minimum of $\hat{N}(\mu)$ is achieved at $\mu = 2$ which corresponds to $m = 1$. We see therefore that the greatest efficiency, assuming ϵ is small compared with $\frac{1}{n}$, is achieved with the least possible number of points per block. This dictates $n = 2$ or $n = 3$ and, as we showed earlier, the efficiencies of these two are more or less the same.

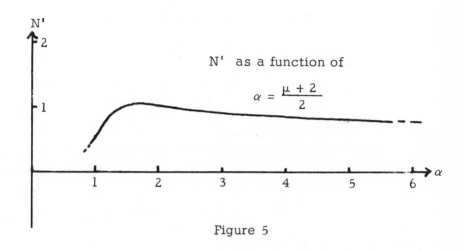

Figure 5

In practice $n = 3$ is to be preferred to $n = 2$. The
optimal placement for $n = 2$ places the points r_1 and r_2
very close to each other, making discrimination between $f(r_1)$
and $f(r_2)$ quite difficult. For $n = 3$ all points are evenly
spaced and this problem is avoided. For similar reasons
$n = 5$ is preferable to $n = 4$, $n = 7$ is preferable to $n = 6$,
and so on, i. e. , odd-block search is best and 3-block search
is best of all. We emphasize, however, that this judgment
is based on our assumption that ϵ is a very small tolerance
since our various approximations are not otherwise valid. Our
approximations have not led us too far astray, however. Tak-
ing $\epsilon = \frac{1}{50}$, $b - a = 1$, we compare 3-blocks with 5-blocks.
For 3-blocks, $L = \frac{2}{3+1} = \frac{1}{2}$. We want $(\frac{1}{2})^k \leq \frac{1}{25}$ which means

k = 5. The number of function evaluations is $3 + 2 + 2 + 2 +$

$2 = 11$. For 5-blocks we have $L = \dfrac{2}{5+1} = \dfrac{1}{3}$. We want $(\dfrac{1}{3})^k$

$\leq \dfrac{1}{25}$ which means k = 3. The number of function evalua-

tions is $5 + 4 + 4 = 13$. Thus, even here, 3-blocks are best

among the odd-block search techniques. As noted earlier,

2-blocks are slightly better, requiring only 10 evaluations if

δ is negligible. But we have also noted that making δ small

creates its own difficulties.

The question of which method should actually be used

cannot be answered without more knowledge of the problem at

hand. If we are seeking the maximum of some well defined

function f expressed in terms of elementary functions such

as x^p, e^x, sin x, etc. , then chances are we would wish to

minimize the total number of function evaluations and hence

we should use a small value of n such as n = 3. In fact

we will soon see that the search technique of greatest effi-

ciency from this point of view involves one point per block.

But the total number of function evaluations is not always the

sole criterion when it comes to judging the effectiveness of a

search technique.

EXAMPLE. Let us assume that a certain chemical process requires eight hours for completion. Moreover we will assume that the initial conditions can be arranged precisely enough so that the evolution of the process is essentially the same for any such run. The objective is to determine the maximum temperature achieved in the chemical mixture and how much time elapses from the beginning of the reaction until this maximum temperature is achieved; this time interval to be measured with a tolerance of one minute, + or -.

We will assume that each experiment costs \$100 to run and that each temperature measurement costs \$5. We will assume that odd-block search is to be used. Now the final interval $[a_k, b_k]$ must have a length not exceeding two minutes if we are to specify the time of maximum temperature within a tolerance of one minute. Thus we need $[a_k, b_k] \leq \frac{1}{240} [a_0, b_0]$, i.e., we need $L^k \leq \frac{1}{240}$. Consider then the following table

n	L	k	Cost in dollars
3	$\frac{1}{2}$	8	$885
5	$\frac{1}{3}$	5	$605
7	$\frac{1}{4}$	4	$525
9	$\frac{1}{5}$	4	$565
11	$\frac{1}{6}$	4	$605
13	$\frac{1}{7}$	3	$485
15	$\frac{1}{8}$	3	$515
...	...	...	...
31	$\frac{1}{16}$	2	$505
...	...	...	...
239	$\frac{1}{240}$	1	$1295

The missing entries in the table correspond to increasing values of n which yield no reduction in k, as shown by n = 7, 9, 11, etc. As the cost always increases if n is increased while k is kept fixed there is no need to list these. The cost is computed via the formula

$$\text{Cost} = 5(n + (n - 1)(k - 1)) + 100k.$$

Thus for n = 13, Cost = $5(13 + (12)(2)) + 100(3) = 5(37) + 300 =$

$485. It is clear that this is the least expensive way to find the time of maximum temperature. We run the experiment three times and take thirteen measurements the first time, twelve measurements the second time and twelve measurements the third time.

Any process which is expensive to run or for some other reason difficult to repeat may cause one to prefer a relatively large value of n as in the above example.

It is time now for us to treat the problem of <u>resolution</u>. Let us suppose that we are seeking the maximum of a function f in an interval $[a, b]$. We will assume that we have a limited ability to actually evaluate $f(r)$ at any point r, that is, in attempting to evaluate $f(r)$ all we can be certain of doing is obtaining a number in, say, the interval $[f(r) - \frac{\epsilon}{2}, f(r) + \frac{\epsilon}{2}]$. For example, if we express $f(r)$ as a decimal with three places after the decimal point then $\epsilon = 10^{-3}$. Thus we cannot tell for certain whether or not $f(r_1)$ is different from $f(r_2)$ unless $|f(r_1) - f(r_2)| > \epsilon$. This is a serious matter in the application of search techniques for if $f(r_{k_0})$ is the maximum of the numbers $f(r_1), f(r_2), \ldots, f(r_n)$ it is quite important that we be able to tell whether or not $f(r_{k_0})$ differs from $f(r_{k_0+1})$

and $f(r_{k_0-1})$. If the latter two numbers are both less than

$f(r_{k_0})$ the maximum of the function f may occur anywhere in

the interval (r_{k_0-1}, r_{k_0+1}). If the defects in our evaluation

procedure cause us to conclude that $f(r_{k_0}) = f(r_{k_0-1})$, say,

then we conclude that the maximum lies in $[r_{k_0-1}, r_{k_0}]$,

which may be false.

A dramatic example is provided by the function $f(r) =$

$r^3 - r^4$. The derivative $f'(r) = 3r^2 - 4r^3 = r^2(3 - 4r)$ is non-

negative for $4r \leq 3$, i. e. , $r \leq \frac{3}{4}$, and negative for $r > \frac{3}{4}$ so

that the function $f(r)$ is unimodal on any interval and has a

unique maximum at $r = \frac{3}{4}$. Suppose we start with $[a_0, b_0] =$

$[-1, 1]$ and use diblock $(n = 2)$ search with $\delta = .05$ and

calculate $f(r)$ to three decimal places. In our first block we

have $r_1 = -.05$, $r_2 = .05$ so $f(r_1) = (-.05)^3 - (.05)^4 =$

$-.000125 - .00000625 = .000$ while $f(r_2) = .000125 -$

$.00000625 = .000$ also. We conclude that $f(r_1) = f(r_2)$ so

that the maximum r^* lies in $[-.05, .05]$. Since all suc-

ceeding intervals will be subsets of this one there is no hope

that we will ever find the correct $r^* = \frac{3}{4}$.

Now let r_1 and r_2 be any two points in $[a, b]$. If

r^* lies between r_1 and r_2 it is unimportant whether we dis-

tinguish $f(r_1)$ from $f(r_2)$. Thus we need only consider the

cases where both r_1 and r_2 are less than or equal to r^* and

where r_1 and r_2 are both greater than or equal to r^*.

DEFINITION 3. 2.   Let f be (+) unimodal on [a, b]   with max-

imum at $r^* \in [a, b]$. For a given evaluation tolerance $\epsilon/2$

we define the resolution δ_ϵ by

$$\delta_\epsilon = \text{g. l. b.} \ \{d > 0 \mid |f(r_1) - f(r_2)| > \epsilon \quad \text{whenever}$$

$$|r_1 - r_2| > d \ \text{ and } \ r_1, r_2 \leq r^* \ \text{ or } \ r_1, r_2 \geq r^* \}.$$

If we take $f(r) = -r^2$ then $\delta_\epsilon = \sqrt{\epsilon}$. To see this

assume $r_1 \leq r_2 \leq 0$ and compute $f(r_2) - f(r_1) = -r_2^2 + r_1^2 =$

$(r_1 + r_2)(r_1 - r_2) \geq (r_1 + r_2)(r_1 - r_2) - 2r_2(r_1 - r_2) = (r_1 - r_2)^2$

so that $f(r_2) - f(r_1) > \epsilon$ if $r_2 - r_1 > \sqrt{\epsilon}$. On the other hand

if $r_2 = 0$ then $f(r_2) - f(r_1) = (r_1 - r_2)^2$ so $f(r_2 - f(r_1)$ is

not $> \epsilon$ in general unless $r_2 - r_1 > \sqrt{\epsilon}$. A similar analysis

may be applied to $0 \leq r_1 < r_2$ and we conclude $\delta_\epsilon = \sqrt{\epsilon}$.

In general we cannot hope to determine r^* with error

less than δ_ϵ if our evaluation procedure involves a tolerance

$\epsilon/2$. For if r_1 and r_2 both lie in the interval $[r^* - \delta_\epsilon$, $r^* + \delta_\epsilon]$ it will in general be impossible to distinguish $f(r_1)$ from $f(r_2)$ making further refinement impossible.

The resolution problem definitely affects the usefulness of even-block search techniques. In order to avoid false identification of $f(r_{2k-1})$ with $f(r_{2k})$ it is necessary to take the number δ of Theorem 3.3 greater than $\delta_\epsilon/2$. In actual practice we do not know what δ_ϵ is so we have to be reasonably generous in picking δ and this tends to increase L. In fact we cannot maintain L constant - it must increase with each successive block. This is a powerful argument in favor of odd-block methods. Here there are no small intervals initially and we can proceed with reasonable confidence to employ successive blocks to reduce the interval of uncertainty. A fairly sensible empirical rule is to stop when three successive values of $f(r_k)$ are indistinguishable. In some cases more care than this will be required but such cases are unusual. Generally speaking the problem of resolution is one to be kept in mind but it is probably not worthwhile to spend a great deal of time and effort worrying about it unless prior knowledge indicates a particularly troublesome situation.

We now pass from a discussion of poly-block, $n \geq 2$,

search techniques to what are called uni-block techniques.

Uni-block techniques produce a new interval $[a_k, b_k]$ with

each new point of evaluation of f after the first.

The idea of uni-block search is as follows. We place

two points, r_0^1 and r_0^2 in the interior of the interval $[a, b] =$

$[a_0, b_0]$. Assuming f to be (+) unimodal, we evaluate

$f(r_0^1)$ and $f(r_0^2)$ and choose $[a_1, b_1] = [a, r_0^2]$ if $f(r_0^1) >$

$f(r_0^2)$, $[a_1, b_1] = [r_0^1, b]$ if $f(r_0^2) > f(r_0^1)$ or $[a_1, b_1] = [r_0^1, r_0^2]$

if $f(r_0^1) = f(r_0^2)$. The last of these possibilities is non-typical

so we will concentrate on the first two. If $[a_1, b_1] = [a, r_0^2]$

then the point r_0^1 already lies in the interior of $[a_1, b_1]$.

Therefore we need only add one new point in order to have

two points in the interior of $[a_1, b_1]$ with which to deter-

mine $[a_2, b_2]$. Since $f(r_0^1)$ has already been evaluated,

only one new evaluation of f will be required to pass from

$[a_1, b_1]$ to $[a_2, b_2]$. The situation is the same if $[a_1, b_1] =$

$[r_0^1, b]$ except that now it is r_0^2 which lies in the interior

of $[a_1, b_1]$ already.

At the kth step $[a_k, b_k]$ is cut out of $[a_{k-1}, b_{k-1}]$

using two points r_{k-1}^1, r_{k-1}^2. One of these points is new, one is old. That is, either r_{k-1}^1 or r_{k-1}^2 will be equal to one of the points r_{k-2}^1 or r_{k-2}^2.

EXAMPLE. Let $f(r) = \min\{-r^2 + 4r - 2,\ 4\log_{10} r\}$ in the interval $1 \leq r \leq 4$. Both $-r^2 + 4r - 2$ and $\log_{10} r$ are (+) unimodal in this interval and therefore, as the student is asked to show in one of the exercises at the end of this chapter, $f(r)$ is also (+) unimodal there.

We begin by putting $r_0^1 = 2$, $r_0^2 = 3$.

Step 1.

	$r_0^1 = 2$	$r_2^0 = 3$
$-r^2 + 4r - 2$	2.000	1.000
$4\log_{10} r$	1.204	1.908
$f(r)$	1.204	1.000

Thus $[a_1, b_1] = [1, 3]$ which already contains $r_0^1 = 2$. We take $r_1^1 = 2$ and add one new point $r_1^2 = 2.5$.

Step 2.

	$r_1^1 = r_0^1 = 2$	$r_1^2 = 2.5$
$-r^2 + 4r - 2$	Computed in Step 1	1.750
$4\log r$	Step 1	1.592
$f(r)$	1.204	1.592

Thus $[a_2, b_2] = [2, 3]$ which already includes $r_1^2 = 2.5$.

We take $r_2^2 = 2.5$ and add a new point $r_2^1 = 2.25$.

Step 3.

	$r_2^1 = 2.25$	$r_2^2 = r_1^2 = 2.5$
$-r^2 + 4r - 2$	1.937	Computed
$4 \log r$	1.408	in Step 2
$f(r)$	1.408	1.592

Thus $[a_3, b_3] = [2.25, 3]$ which already includes $r_2^2 = 2.5$.

We take $r_3^1 = r_2^2 = 2.5$ and add a new point $r_3^2 = 2.75$.

Step 4.

	$r_3^1 = r_2^2 = 2.5$	$r_3^2 = 2.75$
$-r^2 + 4r - 2$	Computed	1.437
$4 \log r$	in Step 3	1.756
$f(r)$	1.592	1.437

Thus $[a_4, b_4] = [2.25, 2.75]$. This is as far as we will

carry this example.

The reader will note the rather haphazard manner in

which the new points were chosen in the above example. When-

ever possible we chose the new point so that r_k^1, r_k^2 were

symmetrically placed in $[a_k, b_k]$. However, this was

impossible whenever the old point lay at the center of the interval $[a_k, b_k]$. When this happens the choice of the new
point is more or less arbitrary. This arbitrariness is unattrac-
tive and, as can be shown, wasteful. Also, the ratio of the
length of $[a_{k+1}, b_{k+1}]$ to $[a_k, b_k]$ does not remain fixed
in this example, which would make it difficult to determine
in advance how many steps would be required to obtain r^*
within a given tolerance.

Let us place two points, r_0^1, r_0^2 in an interval $[a, b]$
so that they are located symmetrically. That is, we put $r_0^2 =$
$a + L(b - a)$, $r_0^1 = b - L(b - a)$, $\frac{1}{2} < L < 1$. Then the ratio of
the length of $[a_1, b_1]$ to the length of $[a, b]$ is L, no
matter whether $[a_1, b_1] = [a, r_0^2]$ or $[a_1, b_1] = [r_0^1, b]$.
Since there two situations are completely symmetric, let us
assume that $[a_1, b_1] = [a, r_0^2]$. Now we wish to place a
new point $\tilde{r}$ in $[a, r_0^2]$ so that r_0^1 and $\tilde{r}$ are symmetric in
that interval. Since $r_0^1 = a + (1 - L)(b - a)$ we want $\tilde{r} =$
$r_0^2 - (1 - L)(b - a)$. If this is done, the length of $[a_2, b_2]$
will be $r_0^1 - a = r_0^2 - \tilde{r}$, provided it turns out that $\tilde{r} < r_0^1$,
as shown in Figure 6.

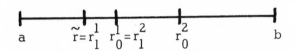

Figure 6

Now $r_0^1 - a = (1 - L)(b - a)$. Thus

$$b_0 - a_0 = b - a$$

$$b_1 - a_1 = L(b - a)$$

$$b_2 - a_2 = (1 - L)(b - a).$$

We would like the ratio of the length of $[a_2, b_2]$ to the length of $[a_1, b_1]$ to be the same as the ratio of the length of $[a_1, b_1]$ to the length of $[a_0, b_0]$. Thus we want

$$\frac{1 - L}{L} = L, \qquad L^2 + L - 1 = 0.$$

Using the quadratic formula

$$L = \frac{-1 \pm \sqrt{1 + 4}}{2} = \frac{-1 \pm \sqrt{5}}{2}.$$

As L must be positive we take $L = \frac{\sqrt{5}-1}{2} = .618034\ldots$

THEOREM 3.4. Let the points r_0^1, r_0^2 be chosen in $[a, b] = [a_0, b_0]$ in such a way that

$$r_0^2 = a + \frac{\sqrt{5}-1}{2}(b-a), \qquad r_0^1 = b - \frac{\sqrt{5}-1}{2}(b-a.$$

Then in successive intervals $[a_k, b_k]$ it will always be possible to place the points r_k^1, r_k^2 symmetrically and, if this is done, we shall have

$$r_k^2 = a_k + \frac{\sqrt{5}-1}{2}(b_k - a_k), \qquad r_k^1 = b_k - \frac{\sqrt{5}-1}{2}(b_k - a_k)$$

and thus the ratio of the length of $[a_{k+1}, b_{k+1}]$ to that of $[a_k, b_k]$ will always be $\frac{\sqrt{5}-1}{2}$.

Proof. The truth of our statements is easily verified for $k = 1, 2$. In order to obtain a proof by mathematical induction let us assume these statements are true for $k \leq k_0 - 1$. Suppose it turns out that $[a_{k_0}, b_{k_0}] = [a_{k_0-1}, r_{k_0-1}^2]$. Since

$$r_{k_0-1}^2 = a_{k_0-1} + \frac{\sqrt{5}-1}{2}(b_{k_0-1} - a_{k_0-1})$$

$$r^1_{k_0-1} = b_{k_0-1} - \frac{\sqrt{5}-1}{2}(b_{k_0-1} - a_{k_0-1})$$

$$r^2_{k_0-1} - r^1_{k_0-1} = (\sqrt{5}-2)(b_{k_0-1} - a_{k_0-1}).$$ Since

$\sqrt{5} - 2 < \frac{1}{2}(\frac{\sqrt{5}-1}{2})$ we take $r^2_{k_0} = r^1_{k_0-1}$ and we can place

$r^1_{k_0}$ symmetrically at

$$r^1_{k_0} = a_{k_0-1} + (\sqrt{5}-2)(b_{k_0-1} - a_{k_0-1})$$

$$= a_{k_0} + (\sqrt{5}-2)(\frac{2}{\sqrt{5}-1})(b_{k_0} - a_{k_0})$$

$$= a_{k_0} + (2 - \frac{2}{\sqrt{5}-1})(b_{k_0} - a_{k_0})$$

$$= b_{k_0} + ((2 - \frac{2}{\sqrt{5}-1}) - 1)(b_{k_0} - a_{k_0})$$

$$= b_{k_0} + (\frac{\sqrt{5}-1-2}{\sqrt{5}-1})(b_{k_0} - a_{k_0})$$

$$= b_{k_0} - (\frac{\sqrt{5}-1}{2})(b_{k_0} - a_{k_0}).$$

Similarly $r^2_{k_0} = a_{k_0} + (\frac{\sqrt{5}-1}{2})(b_{k_0} - a_{k_0})$. The calculations

are entirely the same if it turns out that $[a_{k_0}, b_{k_0}] = [r^1_{k_0-1}, b_{k_0-1}]$. Thus we may regard the proof as complete.

The formulas of Theorem 3.4, which may be rewritten

$$r^2_k = \frac{3-\sqrt{5}}{2}a_k + \frac{\sqrt{5}-1}{2}b_k, \quad r^1_k = \frac{\sqrt{5}-1}{2}a_k + \frac{3-\sqrt{5}}{2}b_k$$

serve to generate a uni-block search scheme known as

Golden Section search. The name is due to the fact that a

rectangle whose sides are in the ratio $1 : \frac{\sqrt{5}-1}{2}$ has often

been considered ideally proportioned from the artistic point

of view. For example, a standard $8" \times 5"$ picture frame has

roughly these proportions.

        We now give an example of the use of Golden Section

search.

EXAMPLE. We again take $f(r) = \min = \{-r^2 + 4r - 2, \ 4 \log_{10} r\}$

in the interval $[1, 4] = [a_0, b_0]$. Working to three decimal

places we have $\frac{\sqrt{5}-1}{2} = .618$ and hence $r_0^1 = 4 - .618(4-1) =$

$4 - 1.854 = 2.146$ while $r_0^2 = 1 + .618(4-1) = 2.854$.

Step 1.

	$r_0^1 = 2.146$	$r_0^2 = 2.854$
$-r^2 + 4r - 2$	1.979	1.271
$4 \log_{10} r$	1.326	1.823
$f(r)$	1.326	1.271

The new interval $[a_1, b_1] = [1, 2.854]$ which already con-

tains $r_1^2 = 2.146$. Noting that $r_1^1 - a_1 = b_1 - r_1^2 = 2.854 -$

2.146 we have $r_1^1 = 1 + .708 = 1.708$.

Step 2.

	$r_1^1 = 1.708$	$r_1^2 = 2.146$
$-r^2 + 4r - 2$	1.915	Already computed in
$4 \log_{10} r$	.930	Step 1
$f(r)$	.930	1.326

Thus $[a_2, b_2] = [1.708, 2.854]$ which already includes $r_1^2 = 2.146$. Then, since $b_2 - r_2^2 = r_2^1 - a_2 = 2.146 - 1.708 = .438$ we have $r_2^2 = 2.854 - .438 = 2.416$.

Step 3.

	$r_2^1 = 2.146$	$r_2^2 = 2.416$
$-r^2 + 4r - 2$	Already computed in	1.827
$4 \log_{10} r$	Step 2	1.528
$f(r)$	1.326	1.528

We have $[a_3, b_3] = [2.146, 2.854]$ which already includes $r_3^1 = 2.416$. Since $b_3 - r_3^2 = r_3^1 - a_3 = 2.416 - 2.146 = .270$ we have $r_3^2 = b_3 - .270 = 2.854 - .270 = 2.584$.

Step 4.

	$r_3^1 = 2.416$	$r_3^2 = 2.584$
$-r^2 + 4r - 2$	Already computed in	1.659
$4 \log_{10} r$	Step 3	1.648
$f(r)$	1.528	1.648

We have $[a_4, b_4] = [a.146, 2.854]$ which already includes
$r_4^1 = 2.584$. Since $b_4 - r_4^2 = r_4^1 - a_4 = 2.584 - 2.146 = .168$,
$r_4^2 = b_4 - .168 = 2.854 - .168 = 2.686$.

Step 5.

	$r_4^1 = 2.584$	$r_4^2 = 2.686$
$-r^2 + 4r - 2$	Already computed in Step 4	1.530
$4 \log_{10} r$		1.716
$f(r)$	1.648	1.530

This yields $[a_5, b_5] = [2.416, 2.686]$. We will stop at
this point, since the search pattern has been clearly demon-
strated.

According to Theorem 3.4 we always have

$$(b_{k+1} - a_{k+1}) = \frac{\sqrt{5} - 1}{2}(b_k - a_k)$$

and from this it readily follows that for all k

$$(b_k - a_k) = (\frac{\sqrt{5} - 1}{2})^k (b_0 - a_0) = (\frac{\sqrt{5} - 1}{2})^k (b - a).$$

If we need to know r^* with a tolerance ϵ we must have
$b_k - a_k \leq 2\epsilon$ and so we need

$$(\frac{\sqrt{5} - 1}{2})^k (b - a) \leq 2\epsilon.$$

Taking logarithms to base 10

$$k \log_{10} \left(\frac{\sqrt{5} - 1}{2}\right) + \log_{10} (b - a) \leq \log_{10} 2 + \log_{10} \epsilon$$

whence, since $\log_{10} \left(\frac{\sqrt{5} - 1}{2}\right)$ is negative,

$$k \geq \frac{\log_{10} 2 + \log_{10} \epsilon - \log_{10} (b - a)}{\log_{10} \left(\frac{\sqrt{5} - 1}{2}\right)}$$

Given ϵ, the number of steps required is the smallest positive integer k which satisfies this inequality. Suppose in the example $f(r) = \min\{-r^2 + 4r - 2, \ 4 \log_{10} r\}$ studied above we wished $\epsilon = .001$. Taking $\log_{10} \left(\frac{\sqrt{5} - 1}{2}\right) = -.209$ we need

$$k \geq \frac{.301 - 3.000 - .477}{-.209} = \frac{3.176}{.209} = 15.19$$

and the required number of steps is seen to be 16.

Let us compare the efficiency of Golden Section search with tri-block search. For tri-block search $(b_k - a_k) = \left(\frac{1}{2}\right)^k (b - a)$ and therefore to obtain r^* within a tolerance ϵ we need

$$k \geq \frac{\log_{10} 2 + \log_{10} \epsilon - \log_{10} (b - a)}{\log_{10} \left(\frac{1}{2}\right)} \ .$$

Comparing this with the similar expression obtained above

for Golden Section search we see that if k_3 represents the

number of steps in tri-block search and k_1 the number of

steps in Golden Section search then, approximately,

$$\frac{k_1}{k_3} = \frac{\log_{10}(\frac{1}{2})}{\log_{10}(\frac{\sqrt{5}-1}{2})} \; .$$

But tri-block search requires two new points at each step

while Golden Section search requires only one new point at

each step. Thus if n_1 and n_3 represent the total number

of function evaluations then, again approximately,

$$\frac{n_1}{n_3} = \frac{k_1}{2k_3} = \frac{\frac{1}{2}\log_{10}(\frac{1}{2})}{\log_{10}(\frac{\sqrt{5}-1}{2})} = \frac{-.151}{-.209} = .723$$

so that Golden Section search requires the evaluation of $f(r)$

at about $3/4$ as many points as does tri-block search and

is thus somewhat more efficient from this point of view.

In Golden Section search all points r_k^j, $j = 1, 2,$

$k = 1, 2, 3, \ldots$ depend very critically upon r_0^1 and r_0^2.

Therefore r_0^1 and r_0^2 should be computed very accurately in

accordance with Theorem 3.4. This means that $\frac{\sqrt{5}-1}{2}$ should

be carried out to as many decimal or binary places as the
computer will allow.

EXERCISES.   CHAPTER 3

1.   Let $f_1, f_2, \ldots, f_n$ be functions, $f_i : R^1 \to R^1$.   Define

$$f(r) = \min_{i = 1, 2, \ldots, n} \{f_i(r)\}, \quad r \in R^1.$$

Show that if the functions f_i are all (+) unimodal on an
interval $[a, b]$ then f is also (+) unimodal there.
(Hint: show the result is true for $n = 2$ and then use
mathematical induction on n.)

2.   Let f be a function defined on an interval $[a, b]$ whose
behavior is <u>simple</u> in the following sense: it is possible
to write $[a, b] = [a, c_1] \cup [c_1, c_2] \cup \ldots \cup [c_n, b]$, where
$a < c_1 < c_2 < \ldots < c_n < b$, and f is either (+) or (-)
unimodal in each of these intervals.   What will be the
result if we apply di-block (n = 2) search to such a
function in an effort to find a maximum?   Assume case
(ii) of Theorem 3.1 never arises.

3. Combine the result of problem 1 with a tri-block search technique to solve the equation $\log_e x = e^{-x} + 1$, correct to three decimals.

4. An n-block search technique is to be used to find $\max_{r \epsilon [0,1]} f(r)$ with a tolerance .0001. How should we choose n so as to minimize the sum of the number of function evaluations and the number of blocks used?

5. Assume di-block search is to be used to find $\max_{r \epsilon [0,1]} f(r)$, where $f(r)$ is (+) unimodal with resolution δ_ϵ = .001. How many blocks will be required to find the maximum within .0015? Obviously you cannot ignore the resolution δ_ϵ in computing your answer.

6. Write a computer program which employs penta-block (n = 5) search to find the minimum of a (−) unimodal function f on an interval [a, b] within a given tolerance. Be certain that f is never evaluated twice at the same point. Test your routine on

    [a, b] = [0, 5] ,

    tolerance = .001 ,

    $f(r) = e^x - 5x$.

7.   Repeat problem 6 for Golden Section uni-block search.

8.   Can you devise any uni-block search scheme other than

Golden Section search which will place the points of

evaluation r_k^j in a regular manner? By "regular" we

mean in such a way that the lengths of the successive

intervals $[a_k, b_k]$ form a sequence with a pattern suf-

ficiently simple so that one can compute the number of

steps required for a given tolerance ϵ and starting in-

terval $[a, b]$.

CHAPTER 4.   LINEAR ALGEBRA AND THE LEAST SQUARES PROBLEM

We begin this chapter by studying a certain class of functions called <u>linear</u> functions.

DEFINITION 4. 1.   A function $f : R^n \to R^m$ is called a <u>linear function</u> if whenever x_1 and x_2 are vectors in R^n and α_1 and α_2 are real numbers we have

$$f(\alpha_1 x_1 + \alpha_2 x_2) = \alpha_1 f(x_1) + \alpha_2 f(x_2).$$

When $m = 1$ such a function is often called a <u>linear functional.</u> When $m > 1$ it is often called a <u>linear transformation.</u>

We wish to develop now a convenient notation, i. e. , matrix notation, for describing such linear functions.   In the n dimensional space R^n we distinguish n vectors

$e_1, e_2, \ldots, e_n$. The vector e_i has components

$$e_i^j = \begin{cases} 1 & \text{if } j = i \\ \\ 0 & \text{if } j \neq i, \quad i = 1, 2, \ldots, n, \quad j = 1, 2, \ldots, n. \end{cases}$$

These vectors constitute a <u>basis</u> for R^n, which means that each vector $x \in R^n$ can be written as a unique linear combination of the vectors e_i. In fact, if $x = \begin{pmatrix} x^1 \\ x^2 \\ \vdots \\ x^n \end{pmatrix}$ we have

$x = x^1 e_1 + x^2 e_2 + \ldots + x^n e_n$. This may be verified without difficulty and it is also clear that there are no coefficients $\xi^1, \xi^2, \ldots, \xi^n$ different from $x^1, x^2, \ldots, x^n$ for which $x = \xi^1 e_1 + \xi^2 e_2 + \ldots + \xi^n e_n$. In the space R^m we do the same thing. We distinguish m vectors $\hat{e}_i$, $i = 1, 2, \ldots, m$, with

$$\hat{e}_i^j = \begin{cases} 1 & \text{if } j = i \\ \\ 0 & \text{if } j \neq i, \quad i = 1, 2, \ldots, m, \quad j = 1, 2, \ldots, m. \end{cases}$$

If $y \in R^n$, then $y = y^1 e_1 + y^2 e_2 + \ldots + y^m e_m$.

If $f : R^n \to R^m$ is a linear function then Definition 4.1 implies right away that for each $x \in R^n$

$$f(x) = f(x^1 e_1 + x^2 e_2 + \ldots + x^n e_n)$$

$$= x^1 f(e_1) + x^2 f(e_2) + \ldots + x^n f(e_n).$$

Now each of the vectors $f(e_i)$ lies in R^m, since $f: R^n \to R^m$. Consequently, each vector $f(e_i)$ can be uniquely written as a linear combination of the vectors $\hat{e}_1, \hat{e}_2, \ldots, \hat{e}_m$, i. e.

$$f(\hat{e}_i) = f_i^1 \hat{e}_1 + f_i^2 \hat{e}_2 + \ldots + f_i^m \hat{e}_m.$$

The numbers f_i^j, $j = 1, 2, \ldots, m$, are the components of the vector $f(e_i)$, i. e. $(f(e_i))^j = f_i^j$.

Let us now ask, what is the j-th component $(f(x))^j$ of the m-vector $f(x)$? Clearly

$$(f(x))^j = (x^1 f(e_1) + x^2 f(e_2) + \ldots + x^n f(e_n))^j$$

$$= x^1 (f(e_1))^j + x^2 (f(e_2))^j + \ldots + x^n (f(e_n))^j$$

$$= x^1 f_1^j + x^2 f_2^j + \ldots + x^n f_n^j.$$

That is, if we denote $f(x)$ by y,

$$y^j = \sum_{i=1}^{n} f_i^j x^i.$$

It is clear now that if we know all of the numbers f_i^j, $i = 1, 2, \ldots, n$, $j = 1, 2, \ldots, m$, then we have a complete description of the linear function f. For, given any $x \in R^n$, we can readily compute $y = f(x)$ by means of the m equations

$$y^1 = f_1^1 x^1 + f_2^1 x^2 + \ldots + f_n^1 x^n$$

$$y^2 = f_1^2 x^1 + f_2^2 x^2 + \ldots + f_n^2 x^n$$

$$\vdots \qquad \vdots \qquad \vdots \qquad \qquad \vdots$$

$$y^m = f_1^m x^1 + f_2^m x^2 + \ldots + f_n^m x^n \ .$$

It is customary to write this set of equations in the form

$$\begin{pmatrix} y^1 \\ y^2 \\ \vdots \\ y^m \end{pmatrix} = \begin{pmatrix} f_1^1 & f_2^1 & \cdots & f_n^1 \\ f_1^2 & f_2^2 & \cdots & f_n^2 \\ \vdots & \vdots & & \vdots \\ f_1^m & f_2^m & \cdots & f_n^m \end{pmatrix} \begin{pmatrix} x^1 \\ x^2 \\ \vdots \\ x^n \end{pmatrix}$$

Thus the j-th component of the vector y on the left is computed by using the numbers f_i^j from the j-th row of the rectangular array of numbers shown together with the components of x. The rectangular array of numbers f_i^j is called an $m \times n$ matrix, the number m referring to the

number of rows while the number n refers to the number of

columns.   More specifically, this $m \times n$ array is the <u>matrix</u>

of the linear function f with respect to the bases $e_1, e_2, ...,$

e_n and $\hat{e}_1, \hat{e}_2, ..., \hat{e}_m$.   We denote this matrix by F and

write

$$y = Fx.$$

From now on we will usually use this notation rather than

$y = f(x)$ to describe linear functions.

Two cases deserve special attention, those where

$m = 1$ or $n = 1$.   When $n = 1$ we have a function $f: R^1 \to R^m$

$$\begin{pmatrix} y^1 \\ y^2 \\ \vdots \\ y^m \end{pmatrix} = \begin{pmatrix} f_1^1 \\ f_1^2 \\ \vdots \\ f_1^m \end{pmatrix} (x^1).$$

In this case the matrix F reduces to a vector, namely $\begin{pmatrix} f_1^1 \\ f_1^2 \\ \vdots \\ f_1^m \end{pmatrix}$

and the vector y is just a multiple of this vector.   When

$m = 1$ the linear function $y = Fx$ is often called a <u>linear</u>

<u>functional.</u> The equation $y = Fx$ is now just

$$y^1 = (f^1_1 \; f^1_2 \; \cdots \; f^1_n) \begin{pmatrix} x^1 \\ x^2 \\ \vdots \\ x^n \end{pmatrix}$$

which means $y^1 = f^1_1 x^1 + f^1_2 x^2 + \ldots + f^1_n x^n$. The single-
rowed matrix

$$F = (f^1_1 \; f^1_2 \; \cdots \; f^1_n)$$

is called a <u>row vector,</u> as distinguished from the <u>column</u>

<u>vectors</u> which we have discussed up to the present. We will

denote the set of all such n-dimensional row vectors by $(R^n)'$.

Thus

$$(R^n)' = \{(a_1, a_2, \ldots, a_n) | a_i \in R^1, \; i = 1, 2, \ldots, n\}.$$

Note that in the case of row vectors we normally dis-

tinguish different vectors by superscripts and different com-

ponents of the same vector by subscripts - just the opposite

of what we do in the case of column vectors.

If $x = \begin{pmatrix} x^1 \\ x^2 \\ \vdots \\ x^n \end{pmatrix} \in R^n$, we denote by x' the row vector

$$x' = (x^1, x^2, \ldots, x^n).$$

If $a = (a_1, a_2, \ldots, a_n) \in (R^n)'$, then a' is the column vector

$$a' = \begin{pmatrix} a_1 \\ a_2 \\ \vdots \\ a_n \end{pmatrix}.$$

We see then that if $b = \begin{pmatrix} b^1 \\ b^2 \\ \vdots \\ b^n \end{pmatrix}$ is any fixed vector in R^n,

we can construct a linear functional on R^n by writing

$$y = b'x, \qquad x \in R^n,$$

i. e.

$$y = (b^1, b^2, \ldots, b^n) \begin{pmatrix} x^1 \\ x^2 \\ x^3 \\ \vdots \\ x^n \end{pmatrix} = b^1 x^1 + b^2 x^2 + \ldots + b^n x^n .$$

We will use this notation very frequently. Clearly all linear functionals on R^n have this form for some vector $b \in R^n$.

Let f and g be linear functions with the same domain and range spaces, i. e. , $f : R^n \to R^m$ and $g : R^n \to R^m$. If we form a linear combination of these two functions

$$k(x) = \beta_1 f(x) + \beta_2 g(x)$$

then $k : R^n \to R^m$ is still a linear function for $k(\alpha_1 x_1 + \alpha_2 x_2) =$

$\beta_1 f(\alpha_1 x_1 + \alpha_2 x_2) + \beta_2 g(\alpha_1 x_1 + \alpha_2 x_2) = \beta_1 \alpha_1 f(x_1) + \beta_1 \alpha_2 f(x_2) +$

$\beta_2 \alpha_1 g(x_1) + \beta_2 \alpha_2 g(x_2) = \alpha_1 (\beta_1 f(x_1) + \beta_2 g(x_1)) + \alpha_2 (\beta_1 f(x_2) +$

$\beta_2 g(x_2)) = \alpha_1 k(x_1) + \alpha_2 k(x_2)$.

To compute the matrix K associated with the linear function $k(x)$ we compute

$$k(e_i) = \beta_1 f(e_i) + \beta_2 g(e_i)$$

so that

$$(k(e_i))^j = \beta_1 (f(e_i))^j + \beta_2 (g(e_i))^j$$

and it follows that

$$k_i^j = \beta_1 f_i^j + \beta_2 g_i^j .$$

Thus each entry of the matrix K is β_1 times the corresponding entry of F plus β_2 times the corresponding entry of G.

We write

$$K = \beta_1 F + \beta_2 G.$$

Thus if we take the two matrices F and G, both of dimension $m \times n$, and form the linear combination $\beta_1 F + \beta_2 G$, we obtain the matrix of the linear function $k(x) = \beta_1 f(x) + \beta_2 g(x)$.

Consider now three vector spaces, R^n, R^m and R^p and let us suppose we have two functions $f : R^n \to R^m$ and $g : R^m \to R^p$. The <u>composition</u> of these two functions is a function $h : R^n \to R^p$ which is defined by

$$z = h(x) = g(f(x)),$$

i. e. $y = f(x),$

$z = g(y).$

Thus, given $x \in R^n$, x determines a vector $y \in R^m$ via the function f and y in turn determines a further vector $z \in R^p$. The resulting correspondence between vectors $x \in R^n$ and vectors $z \in R^p$ is the function h.

If $f : R^n \to R^m$ and $g : R^m \to R^p$ are both linear functions it is clear that their composition $h : R^n \to R^p$ is also linear. For

$$h(\alpha_1 x_1 + \alpha_2 x_2) = g(f(\alpha_1 x_1 + \alpha_2 x_2))$$

$$= g(\alpha_1 f(x_1) + \alpha_2 f(x_2)) = \alpha_1 g(f(x_1)) + \alpha_2 g(f(x_2))$$

$$= \alpha_1 h(x_1) + \alpha_2 h(x_2).$$

Consequently there must be matrices F, G, H of dimensions m × n, p × m, p × n, respectively, such that $y = f(x)$, $z = g(y)$, $z = h(x)$ correspond to

$$y = Fx, \quad z = Gy, \quad z = Hx.$$

Since F and G determine the functions f and g which in turn determine h which determines H, F and G together determine H. Just what is this relationship? Clearly what we have to do is to compute $(h(e_i))^j$. Now

$$h(e_i) = g(f(e_i)) = g(\sum_{k=1}^{m} f_i^k \hat{e}_k)$$

$$= \sum_{k=1}^{m} f_i^k g(\hat{e}_k) = \sum_{k=1}^{m} f_i^k (\sum_{j=1}^{p} g_k^j \tilde{e}_j)$$

$$= \sum_{j=1}^{p} (\sum_{k=1}^{m} g_k^j f_i^k) \tilde{e}_j$$

where the vectors $\tilde{e}_1, \tilde{e}_2, \ldots, \tilde{e}_p$ in R^p are defined by

$$\tilde{e}_j^{\,\ell} = \begin{cases} 1 & \text{if } \ell = j \\ 0 & \text{if } \ell \neq j, \; j = 1, 2, \ldots, p, \; \ell = 1, 2, \ldots, p. \end{cases}$$

It follows that

$$h_i^j = (h(e_i))^j = \sum_{k=1}^m g_k^j f_i^k \, .$$

Thus h_i^j, the entry in the j-th row and i-th column of the

matrix H, is a sum of products of entries from the j-th row

of G and the i-th column of F. We write

$$H = GF$$

and we say that H is the product of G and F. This pro-

duct is defined if and only if the number of columns of G

agrees with the number of rows of F. In more detail we have

$$\begin{pmatrix} h_1^1 & h_2^1 & \cdots & h_n^1 \\ h_1^2 & h_2^2 & \cdots & h_n^2 \\ \vdots & \vdots & & \vdots \\ h_1^p & h_2^p & \cdots & h_n^p \end{pmatrix} = \begin{pmatrix} g_1^1 & g_2^1 & \cdots & g_m^1 \\ g_1^2 & g_2^2 & \cdots & g_m^2 \\ \vdots & \vdots & & \vdots \\ g_1^p & g_2^p & \cdots & g_m^p \end{pmatrix} \begin{pmatrix} f_1^1 & f_2^1 & \cdots & f_n^1 \\ f_1^2 & f_2^2 & \cdots & f_n^2 \\ \vdots & \vdots & & \vdots \\ f_1^m & f_2^m & \cdots & f_n^m \end{pmatrix}$$

We have underlined the element h_1^2, the second row of G

and the first column of F, indicating that h_1^2 is formed using

the entries in the second row of G and the entries in the

first column of F. Note that the product FG does not in

general equal GF. In fact FG is not even defined unless

$p = n$.

EXAMPLES.

(i)

$$\begin{pmatrix} 2 & 3 & -1 \\ 4 & 6 & 2 \end{pmatrix} \begin{pmatrix} 3 & 0 & 1 \\ 1 & 1 & -1 \\ 0 & 1 & 2 \end{pmatrix} =$$

$$\begin{pmatrix} (2)(3)+(3)(1)+(-1)(0) & (2)(0)+(3)(1)+(-1)(1) & (2)(1)+(3)(-1)+(-1)(2) \\ (4)(3)+(6)(1)+(2)(0) & (4)(0)+(6)(1)+(2)(1) & (4)(1)+(6)(-1)+(2)(2) \end{pmatrix}$$

$$= \begin{pmatrix} 8 & 2 & -3 \\ 18 & 8 & 2 \end{pmatrix}$$

(ii)

$$\begin{pmatrix} 2 & -1 \\ 3 & 4 \end{pmatrix} \begin{pmatrix} 3 & 2 \\ -1 & 1 \end{pmatrix} = \begin{pmatrix} 7 & 3 \\ 5 & 10 \end{pmatrix}$$

(iii)

$$\begin{pmatrix} 3 & 2 \\ -1 & 1 \end{pmatrix} \begin{pmatrix} 2 & -1 \\ 3 & 4 \end{pmatrix} = \begin{pmatrix} 12 & 5 \\ 1 & 5 \end{pmatrix}$$

It should be noted that the definition of the product

of two matrices is consistent with the product of a matrix and

a vector. In fact, if we denote the columns of H by

$h_1, h_2, \ldots, h_n$ so that $H = (h_1, h_2, \ldots, h_n)$, and the columns

of F by column vectors $f_1, f_2, \ldots, f_n$, so that $F = (f_1, f_2, \ldots,$

$f_n)$, then the equation $H = GF$ is the same as

$$h_i = G f_i, \quad i = 1, 2, \ldots, n .$$

If we denote the rows of H by row vectors $h^1, h^2, \ldots, h^p$

so that $H = \begin{pmatrix} h^1 \\ h^2 \\ \vdots \\ h^p \end{pmatrix}$ and the rows of G by $g^1, g^2, \ldots, g^p$ so

that $g = \begin{pmatrix} g^1 \\ g^2 \\ \vdots \\ g^p \end{pmatrix}$ the equation $H = GF$ is the same as

$$h^i = g^i F, \quad i = 1, 2, \ldots, p .$$

Multiplication of matrices can be repeated over and

over, forming products $F_1 F_2 F_3 \ldots F_r$ as long as the number

of columns of F_i agrees with the number of rows of F_{i+1},

$i = 1, 2, \ldots, r-1.$

We turn now to the subject of linear equations. We

do not intend to cover this in great detail as most students

will have had some experience here.   A system of m   linear

equations in n unknowns consists of m equations

$$a_1^1 x^1 + a_2^1 x^2 + \ldots + a_n^1 x^n = y^1$$

$$a_1^2 x^1 + a_2^2 x^2 + \ldots + a_n^2 x^n = y^2$$

$$\vdots \qquad \vdots \qquad \qquad \vdots \qquad \vdots$$

$$a_1^m x^1 + a_2^m x^2 + \ldots + a_n^m x^n = y^m \; ,$$

where the coefficients a_i^j and the right hand sides y^j are

all known a priori.   In matrix notation this becomes

$$Ax = y$$

and the questions which are being asked are these.   Does

there exist a vector $x \in R^n$ such that Ax is equal to the

given vector $y \in R^m$?   Is there only one such x?   How do

we find x ?

The most important case is m = n, where we have n

equations in n unknowns.   Here the matrix A is an n $\times$ n

square array of real numbers.   We assume that the student is

familiar with the notion of the determinant of such an array, which we will denote by the symbol   det(A),   and we remind

him that

(i)   If $y = 0$ (homogeneous equation) the equation $Ax = 0$

   has only the solution $x = 0$ if $\det(A) \neq 0$ while if

   $\det(A) = 0$ there are infinitely many solutions of $Ax = 0$,

   in particular there are non-zero solutions.

(ii)   If $y \neq 0$ (inhomogeneous equation) the equation $Ax = y$

   always has precisely one non-zero solution if $\det(A) \neq 0$.

   If $\det A = 0$ there are vectors y for which no solution

   x exists.

   The treatment of systems of linear equations where

$m \neq n$ will be discussed as needed in the sequel.

DEFINITION 4. 2.   The $n \times n$ matrix

$$
I = \begin{pmatrix} 1 & 0 & \ldots & 0 \\ 0 & 1 & \ldots & 0 \\ \vdots & \vdots & & \vdots \\ 0 & 0 & \ldots & 1 \end{pmatrix}
$$

is called the identity matrix.   It has the property $Ix = x$ for

all $x \in R^n$.

DEFINITION 4. 3.   Let A be an $n \times n$ matrix.   If there

exists an $n \times n$ matrix A^{-1} such that

$$AA^{-1} = I$$

then A^{-1} is called the _inverse of A_ and A is said to be

non-singular.

THEOREM 4. 1.   The $n \times n$ matrix A is non-singular if and

only if $\det(A) \neq 0$.

Proof.   Note that $I = (e_1, e_2, \ldots, e_n)$ and the equation $AB = I$

is the same as

$$Ab_i = e_i, \qquad i = 1, 2, \ldots, n$$

where b_i is the i-th column of B.   If $\det(A) \neq 0$ each of

these equations has a unique solution b_i and hence there is

a unique matrix B such that $AB = I$.   Then $B = A^{-1}$.

On the other hand if $AA^{-1} = I$ then for each $y \in R^n$

$AA^{-1}y = Iy = y$ so that $A^{-1}y$ is a solution of the equation

$Ax = y$.   But if $\det A = 0$ there exist vectors $y \in R^n$ such

that $Ax = y$ has no solution.   Hence if $\det A = 0$ there

cannot be a matrix A^{-1} such that $AA^{-1} = I$.   This completes the proof.

THEOREM 4. 2.   The $n \times n$ matrix A is non-singular if and
only if $Ax = 0$ implies $x = 0$.

Proof. If A is non-singular then, by Theorem 4. 1, det $A \neq 0$,
and if det $A \neq 0$ then $Ax = 0$ has only the zero solution.
On the other hand if $Ax = 0$ has only the zero solution then
det $A \neq 0$, for when det $A = 0$ the homogeneous equation
$Ax = 0$ has infinitely many solutions. If det $A \neq 0$ A is
non-singular and the proof is complete.

  If A and B are both non-singular then so is AB
and $(AB)^{-1} = B^{-1}A^{-1}$. This is more or less obvious, for if
$AA^{-1} = I$ and $BB^{-1} = I$ then $(AB)(B^{-1}A^{-1}) = A(BB^{-1})A^{-1} =$
$A(I)A^{-1} = AA^{-1} = I$. We also note that if A is non-singular
then so is A^{-1} and $(A^{-1})^{-1} = A$. For if $A^{-1}x = 0$ then
$AA^{-1}x = A(0) = 0$. But $AA^{-1}x = Ix = x$ so $x = 0$. Thus
$A^{-1}x = 0$ implies $x = 0$ and, by Theorem 5. 2, we conclude
that A^{-1} is non-singular. Multiplying the equation $AA^{-1} = I$
on the left by A^{-1} and on the right by $(A^{-1})^{-1}$ we have

$$A^{-1}(AA^{-1})(A^{-1})^{-1} = A^{-1}(I)(A^{-1})^{-1} = I$$

which implies that

$$(A^{-1}A)(A^{-1}(A^{-1})^{-1}) = I$$

so that

$$A^{-1}A = I.$$

Thus $A^{-1}A = I$ and $A^{-1}(A^{-1})^{-1} = I$. Letting a_i be the i-th column of A and b_i the i-th column of $(A^{-1})^{-1}$ we have

$$A^{-1}a_i - A^{-1}b_i = e_i - e_i = 0$$

and thus $A^{-1}(a_i - b_i) = 0$ which implies $a_i = b_i$ which shows that $A = (A^{-1})^{-1}$.

The calculation of A^{-1} is quite straightforward but the number of operations required becomes quite large if n is even moderately large. If we let A^{-1} have columns $b_1, b_2, \ldots, b_n$ then the equation

$$AA^{-1} = I$$

is equivalent to

$$Ab_i = e_i, \qquad i = 1, 2, \ldots, n.$$

Since $\det A \neq 0$, each of these equations can be solved for b_i.

EXAMPLE. Calculate A^{-1} where $A = \begin{pmatrix} 4 & 1 & 2 \\ 1 & 6 & 1 \\ 2 & 1 & 10 \end{pmatrix}$. Letting

$A^{-1} = \begin{pmatrix} b_1^1 & b_2^1 & b_3^1 \\ b_1^2 & b_2^2 & b_3^2 \\ b_1^3 & b_2^3 & b_3^3 \end{pmatrix}$ we have

$$4b_1^1 + b_1^2 + 2b_1^3 = 1 \qquad 4b_2^1 + b_2^2 + 2b_2^3 = 0$$

$$b_1^1 + 6b_1^2 + b_1^3 = 0 \qquad b_2^1 + 6b_2^2 + b_2^3 = 1$$

$$2b_1^1 + b_1^2 + 10b_1^3 = 0 \, , \qquad 2b_2^1 + b_2^2 + 10b_2^3 = 0 \, ,$$

$$4b_3^1 + b_3^2 + 2b_3^3 = 0$$

$$b_3^1 + 6b_3^2 + b_3^3 = 0$$

$$2b_3^1 + b_3^2 + 10b_3^3 = 1 \, .$$

Solving these equations we have $A^{-1} = \dfrac{1}{206} \begin{pmatrix} 59 & -8 & -11 \\ -8 & 36 & -2 \\ -11 & -2 & 23 \end{pmatrix}$.

Efficient schemes for performing these computations may be

found in standard textbooks on numerical analysis.

DEFINITION 4.4. Let A be an $m \times n$ matrix, let $x \in R^n$

and $y \in R^n$. Then the product $x'Ay$ is defined and is a 1×1

matrix, i.e., a real number. As we let x and y vary we

we obtain a real valued function of x and y. This function

is called a <u>bilinear form</u> in x and y.

Perhaps the most familiar bilinear form is obtained by

taking $m = n$ and letting A be the $n \times n$ identity matrix.

We then have $x'Iy = x'y$. This quantity is called the <u>inner</u>

<u>product</u> of the vectors x and y. It has the properties

(i) $x'y = y'x$;

(ii) $x'(y_1 + y_2) = x'y_1 + x'y_2$;

(iii) $x'(\alpha y) = \alpha x'y$, α a real number;

(iv) $x'x = \|x\|_e^2$.

The last property is the principal reason why the Euclidean

norm of a vector is the most natural to use in theoretical work.

All these properties are easy to verify directly, from the

formula

$$x'y = x^1 y^1 + x^2 y^2 + \ldots + x^n y^n \, .$$

DEFINITION 4. 5.   Let A be an $m \times n$ matrix with entries

a_i^j, $j = 1, 2, \ldots, m$, $i = 1, 2, \ldots, n$. The $n \times m$ matrix A',

called the <u>transpose of A</u> or the <u>adjoint of A</u>, has entries

$a_i^{'j}$, $j = 1, 2, \ldots, n$, $i = 1, 2, \ldots, m$, defined by

$$a'^j_i = a^i_j .$$

For example, if

$$A = \begin{pmatrix} 2 & -1 \\ 5 & 3 \\ 7 & 1 \end{pmatrix} \quad \text{then} \quad A' = \begin{pmatrix} 2 & 5 & 7 \\ -1 & 3 & 1 \end{pmatrix} .$$

The rows of A become the columns of A' and the columns of A become the rows of A'. Our definition makes it immediately clear that $(A')' = A$.

Let B be an $m \times n$ matrix and A a $p \times m$ matrix. Then AB is a $p \times n$ matrix. We claim that

$$(AB)' = B'A'.$$

First of all we note that the dimensions are right. For A' is an $m \times p$ matrix and B' is an $n \times m$ matrix so that $B'A'$ is an $n \times p$ matrix, as is $(AB)'$. Now let AB, $(AB)'$ and $B'A'$ have entries ξ^j_i, η^i_j and ζ^i_j, respectively, $j = 1, 2, \ldots, p$, $i = 1, 2, \ldots, n$. Then

$$\eta^i_j = \xi^j_i = \sum_{\ell=1}^{m} a^j_\ell b^\ell_i = \sum_{\ell=1}^{m} a'^\ell_j b'^i_\ell$$

$$\zeta_j^i = \sum_{\ell=1}^{m} b'_\ell{}^i a'_j{}^\ell$$

and we conclude $\eta_j^i = \zeta_j^i$, $j = 1, 2, \ldots, p$, $i = 1, 2, \ldots, n$, and hence $(AB)' = B'A'$ as claimed. Similarly $(A_1 A_2 A_3 \cdots A_r)' = A_r' \cdots A_3' A_2' A_1'$.

Now for any $x \in R^m$, $y \in R^n$, $x'Ay$ is a real number, i. e. , a 1×1 matrix, provided A is an $m \times n$ matrix. The transpose of a 1×1 matrix is obviously itself so

$$(x'Ay)' = x'Ay.$$

On the other hand, by the rules just developed

$$(x'Ay)' = y'A'(x')' = y'A'x$$

and hence we have

$$x'Ay = y'A'x,$$

an identity which will frequently prove useful.

DEFINITION 4. 6.   Let A be an $n \times n$ matrix.   Then A is said to be <u>symmetric</u> if $A' = A$.

An example is $A = \begin{pmatrix} 2 & 3 & 1 \\ 3 & 5 & 0 \\ 1 & 0 & -1 \end{pmatrix}$. A symmetric

matrix has the property that entries in positions symmetric

with respect to the main diagonal are equal. Another way of

expressing this is to say that if A is symmetric then the i-th

row of A is the transpose of the i-th column of A. If A is

symmetric and x, y ϵ R^n we have

$$x' Ay = y' A'x = y' Ax,$$

and we say that x' Ay is a <u>symmetric bilinear form.</u>

DEFINITION 4. 7.   Let A be an n × n matrix and let x ϵ R^n.

Then x'Ax is a real number. The real valued function of x

obtained in this way, letting x vary in R^n, is called a

<u>quadratic form in x</u>.

As an example we note that

$$(x^1)^2 + 3x^1 x^2 + 4x^1 x^3 - (x^3)^2$$

is a quadratic form in the vector $x = \begin{pmatrix} x^1 \\ x^2 \\ x^3 \end{pmatrix} \epsilon R^3$.   For

$$(x^1)^2 + 3x^1x^2 + 4x^1x^3 - (x^3)^2 = (x^1,x^2,x^3)\begin{pmatrix} 1 & 3 & 4 \\ 0 & 0 & 0 \\ 0 & 0 & -1 \end{pmatrix}\begin{pmatrix} x^1 \\ x^2 \\ x^3 \end{pmatrix}.$$

If $x'Ax$ is a quadratic form in x we may assume without loss of generality that A is symmetric. To show this we note that $x'Ax = x'A'x$ and hence

$$x'Ax = \tfrac{1}{2}x'Ax + \tfrac{1}{2}x'Ax = \tfrac{1}{2}x'Ax + \tfrac{1}{2}x'A'x$$

$$= x'(\tfrac{1}{2}(A + A'))x.$$

Now $\tfrac{1}{2}(A + A')$ is symmetric, for

$$(A + A')' = A' + (A')' = A' + A = A + A'.$$

Thus $x'Ax$ can always be written $x'(\tfrac{1}{2}(A + A'))x = x'\tilde{A}x$ where $\tilde{A}$ is symmetric. We will normally assume in the future that this has been done when we say that $x'Ax$ is a quadratic form, i. e. , we will assume that A is symmetric. As an example we observe that

$$(x^1,x^2,x^3)\begin{pmatrix} 1 & 3 & 4 \\ 0 & 0 & 0 \\ 0 & 0 & 1 \end{pmatrix}\begin{pmatrix} x^1 \\ x^2 \\ x^3 \end{pmatrix} = (x^1,x^2,x^3)\begin{pmatrix} 1 & \tfrac{3}{2} & 2 \\ \tfrac{3}{2} & 0 & 0 \\ 2 & 0 & -1 \end{pmatrix}\begin{pmatrix} x^1 \\ x^2 \\ x^3 \end{pmatrix}.$$

DEFINITION 4. 8.   The quadratic form $x'Ax$ is said to be

positive definite if $x'Ax > 0$ whenever $x \neq 0$.   If $x'Ax \geq 0$

for all x and there is some $x \neq 0$ for which $x'Ax = 0$ we

say $x'Ax$ is positive semi-definite.   We say $x'Ax$ is posi-

tive if it is either positive definite or positive semi-definite.

The most useful test for positive definiteness is the

following.   If A is symmetric then $x'Ax$ is positive definite

if and only if a_1^1, $\det \begin{pmatrix} a_1^1 & a_2^1 \\ a_1^2 & a_2^2 \end{pmatrix}$, $\det \begin{pmatrix} a_1^1 & a_2^1 & a_3^1 \\ a_1^2 & a_2^2 & a_3^2 \\ a_1^3 & a_2^3 & a_3^3 \end{pmatrix}$ $\ldots,\det(A)$

are all positive.   That is, we form a succession of 1×1,

2×2, $3 \times 3, \ldots, n \times n$ matrices from A in the manner in-

dicated and compute their determinants.   If all of these de-

terminants are positive $x'Ax$ is positive definite, otherwise

it is not.   It is not true that if A is symmetric and all the

above determinants are non-negative then $x'Ax$ is neces-

sarily positive.   We will not prove these results here.

DEFINITION 4. 9.   The $n \times n$ matrix A is said to be positive

definite or positive semi-definite if the quadratic form $x'Ax$

is positive definite or positive semi-definite, respectively.

A is positive if it is either positive definite or positive semi-definite.

To provide examples we note that

$$A_1 = \begin{pmatrix} 4 & 1 & 2 \\ 1 & 6 & 1 \\ 2 & 1 & 10 \end{pmatrix}, \quad A_2 = \begin{pmatrix} 1 & 1 & 1 \\ 1 & 1 & 1 \\ 1 & 1 & 1 \end{pmatrix}$$

are such that A_1 is positive definite while A_2 is positive semidefinite. A_1 is positive definite because $4 > 0$, $\det \begin{pmatrix} 4 & 1 \\ 1 & 6 \end{pmatrix} = 23 > 0$ and $\det A_1 = 214 > 0$. A_2 is positive semidefinite because

$$(x^1, x^2, x^3) \begin{pmatrix} 1 & 1 & 1 \\ 1 & 1 & 1 \\ 1 & 1 & 1 \end{pmatrix} \begin{pmatrix} x^1 \\ x^2 \\ x^3 \end{pmatrix} = (x^1 + x^2 + x^3)^2 \geq 0$$

but vanishes if $x^1 + x^2 + x^3 = 0$. Thus if we take $x = \begin{pmatrix} 2 \\ -1 \\ -1 \end{pmatrix}$ x is not the zero vector but $x'A_2 x = (2 \ -1 \ -1)^2 = 0$.

THEOREM 4.3. If A is a positive definite $n \times n$ matrix then A is nonsingular.

Proof. Suppose $Ax = 0$. Then $x'Ax = x'0 = 0$. Since A is

positive definite we must have $x = 0$. Thus $Ax = 0$ implies $x = 0$ and, by Theorem 4.2, A is non-singular.

THEOREM 4.4. Let A be an $m \times n$ matrix. Then $A'A$ is a positive symmetric matrix. If $Ax = 0$ implies $x = 0$ then $A'A$ is positive definite.

<u>Proof.</u> First of all, $A'A$ is symmetric because $(A'A)' = A'(A')' = A'A$. For any $x \in R^n$ we have $x'A'Ax = (Ax)'(Ax) = \|Ax\|_e^2 \geq 0$. Hence $A'A$ is positive. If $Ax = 0$ implies $x = 0$ then $Ax \neq 0$ whenever $x \neq 0$ and hence $\|Ax\|_e^2 > 0$ when $x \neq 0$. Thus $x'A'Ax > 0$ when $x \neq 0$ and $A'A$ is positive definite. This completes the proof.

THEOREM 4.5. Let A be an $n \times n$ matrix. If A is non-singular then so is A' and $(A')^{-1} = (A^{-1})'$. If A is symmetric and positive definite then A^{-1} is also symmetric and positive definite.

<u>Proof.</u> Since $AA^{-1} = I$, $(AA^{-1})' = I' = I$, i.e.

$$(A^{-1})'A' = I$$

which shows that $((A^{-1})')^{-1} = A'$ and from this, noting the

remarks after Theorem 4.2, we see that $(A')^{-1} = (A^{-1})'$.

To prove the second part we note that if A is symmetric

$$(A^{-1})' = (A')^{-1} = A^{-1}$$

so that A^{-1} is symmetric. Also

$$x'A^{-1}x = x'(A^{-1})AA^{-1}x = x'(A^{-1})'AA^{-1}x$$

$$= (A^{-1}x)'A(A^{-1}x) > 0 \quad \text{if} \quad A^{-1}x \neq 0.$$

Since A is positive definite, hence nonsingular, A^{-1} is non-

singular so $A^{-1}x \neq 0$ when $x \neq 0$. Thus

$$x'A^{-1}x > 0 \quad \text{if} \quad x \neq 0$$

and A^{-1} is positive definite. This completes the proof.

At this point we are, at last, in a position to discuss

the least squares problem. We introduce this topic with an

example.

A surveyor is attempting to determine the height of

each of three mountains, whose heights, we shall assume,

are A, B and C. Standing at sea level he uses various

triangulation techniques, obtaining the results

$$A = 1236 \text{ ft.}, \quad B = 1941 \text{ ft.}, \quad C = 2417 \text{ ft.}$$

Wishing to confirm these measurements he scales the lowest

of these mountains and measures the height of the other two

above the lower, obtaining

$$B - A = 711 \text{ ft.}, \quad C - A = 1177 \text{ ft.}$$

Still not satisfied, he scales the next highest and measures

$$C - B = 475 \text{ ft.}$$

Returning to base the surveyor quickly discovers that his

measurements are inconsistent, though not wildly so. How-

ever, he has no further time and must report his findings.

What heights should he report for the three mountains?

Quite generally in the physical sciences a quantity is

measured a number of times, directly and indirectly. The

measurements thus made almost always contain slight dis-

crepancies due to minor imperfections in the measurement

techniques. From the statistical standpoint it turns out that,

if we assume all measurements equally reliable, it is

reasonable to choose as the final value of this quantity a

number such that the sum of the squares of the differences

between this number and the various measured values of the

physical quantity is as small as possible.

Our surveyor's results may be summarized by a system

of six linear equations in three unknowns:

$$
\begin{pmatrix}
1 & 0 & 0 \\
0 & 1 & 0 \\
0 & 0 & 1 \\
-1 & 1 & 0 \\
-1 & 0 & 1 \\
0 & -1 & 1
\end{pmatrix}
\begin{pmatrix}
A \\
B \\
C
\end{pmatrix}
=
\begin{pmatrix}
1236 \\
1941 \\
2417 \\
711 \\
1177 \\
475
\end{pmatrix} .
$$

These equations are clearly inconsistent. For example, the

first and second equations together give B - A = 705 ft.

while the fourth equation gives B - A = 711 ft. Using the

principle described above for inconsistent measurements, the

surveyor decides to report values A, B and C such that the

quantity

$$(A - 1236)^2 + (B - 1941)^2 + (C - 2417)^2 + (B - A - 711)^2$$

$$+ (C - A - 1177)^2 + (C - B - 475)^2$$

is minimal. Thus he wishes to solve the problem

$$\min_{(A,\,B,\,C)\,\in\,R^3} \left\| \begin{pmatrix} 1 & 0 & 0 \\ 0 & 1 & 0 \\ 0 & 0 & 1 \\ -1 & 1 & 0 \\ -1 & 0 & 1 \\ 0 & -1 & 1 \end{pmatrix} \begin{pmatrix} A \\ B \\ C \end{pmatrix} - \begin{pmatrix} 1236 \\ 1941 \\ 2417 \\ 711 \\ 1177 \\ 475 \end{pmatrix} \right\|_e^2 .$$

Before solving this problem let us abstract it. We consider an equation

$$Ax = b,$$

where x is an n-vector, b is an m-vector and A is an m × n matrix, m ≥ n. When m = n we can, in general, solve this system only if $\det A \neq 0$. When m > n there is no solution unless some special relationship holds among the entries of A and the components of b. It is this last case, m > n, which particularly interests us. Assuming the equation Ax = b has no solution we abandon any attempts to solve it and seek instead for a vector x^* which yields a minimum value for the expression

$$\| Ax - b \|_e^2 .$$

Because this is the same as minimizing

$$\sum_{i=1}^{m} (\sum_{j=1}^{n} a_j^i x^j - b^i)^2$$

we refer to the solution of this new problem, if it exists, as

the _least squares_ solution of $Ax = b$.

THEOREM 4.6. Let A be an $m \times n$ matrix, $m \geq n$, such that

the equation $Ax = 0$ has no non-zero solution. Then there

is a unique vector $x^* \epsilon R^n$ yielding a minimum for $\|Ax-b\|_e^2$

and x^* satisfies

$$A'Ax^* = A'b, \quad \text{i. e.,} \quad x^* = (A'A)^{-1}A'b.$$

Moreover

$$\|Ax^* - b\|_e^2 = b'(I - A(A'A)^{-1}A')b.$$

Proof. The method of proof is essentially that of "completing

the square". Recall that completing the square for a binomial

$a_0 \lambda^2 + a_1 \lambda + a_2$, $a_0 \neq 0$, is accomplished by writing

$$a_0 \lambda^2 + a_1 \lambda + a_2 = a_0(\lambda^2 + \frac{a_1}{a_0}\lambda + \frac{1}{4}(\frac{a_1}{a_0})^2)$$
$$+ a_2 - \frac{1}{4}\frac{a_1^2}{a_0} = a_0(\lambda + \frac{1}{2}\frac{a_1}{a_0})^2 + a_2 - \frac{1}{4}\frac{a_1^2}{a_0}.$$

If $a_0 > 0$ this procedure shows that the binomial achieves its minimum value at $\lambda = -\frac{1}{2}\frac{a_1}{a_0}$ and the minimum value is $a_2 - \frac{1}{4}\frac{a_1^2}{a_0}$.

Since $Ax = 0$ implies $x = 0$, Theorem 4. 5 shows that $A'A$ is positive definite and hence nonsingular, i. e. , $(A'A)^{-1}$ exists. Thus

$$\| Ax - b \|_e^2 = (Ax - b)'(Ax - b) = (x'A' - b')(Ax - b)$$

$$= x'A'Ax - x'A'b - b'Ax + b'b$$

$$= x'A'A(A'A)^{-1}A'Ax - x'A'A(A'A)^{-1}A'b$$

$$- b'A(A'A)^{-1}A'Ax + b'A(A'A)^{-1}A'b$$

$$+ b'(I - A(A'A)^{-1}A')b$$

$$= (A'Ax - A'b)'(A'A)^{-1}(A'Ax - A'b)$$

$$+ b'(I - A(A'A)^{-1}A')b.$$

From Theorem 4. 5 we see that $(A'A)^{-1}$ is positive definite. Therefore

$$(A'Ax - A'b)'(A'A)^{-1}(A'Ax - A'b) \geq 0$$

and equality holds if and only if

$$A'Ax - A'b = 0.$$

Now, since $A'A$ is positive definite, this equation has exactly one solution, namely

$$x^* = (A'A)^{-1}A'b.$$

What we have shown, therefore, is that for all $x \in R^n$

$$\|Ax - b\|_e^2 \geq b'(I - A(A'A)^{-1}A')b$$

and equality holds if and only if $x = x^*$. With this the proof is complete.

The condition that $Ax = 0$ should imply $x = 0$ is easy to check. If A has n rows, say that $i_1, i_2, \ldots, i_n$-th rows, which form a matrix $\tilde{A}$ of dimensions $n \times n$, then this condition holds if $\tilde{A}$ is nonsingular. For the equation

$$Ax = y$$

implies

$$\tilde{A}x = \begin{pmatrix} y^{i_1} \\ \vdots \\ y^{i_n} \end{pmatrix} = \tilde{y} \quad .$$

Since $\tilde{A}$ is nonsingular, $x \neq 0$ implies $\tilde{y} \neq 0$ implies $y \neq 0$. Thus $y = 0$ implies $x = 0$. In the case of the surveyor's problem, e. g. , the first three rows form the 3×3 identity

matrix and our condition is immediately verified.

The quantity $b'(I - A(A'A)^{-1}A')b$ is known as the

residual. It may be considered as a measure of the extent to

which the original equations represented by $Ax = b$ are in-

consistent, being large if the inconsistency is great and small

if the inconsistency is minor.

Note that when $m = n$ the equation $A'Ax^* = A'b$

reduces to $Ax^* = b$. For the condition $Ax = 0$ implies $x = 0$

means A, and hence A' is nonsingular and we can multiply

the equation $A'Ax^* = A'b$ on the left by $(A')^{-1}$ to get $Ax^* =$

b. In this case the residual is easily seen to be zero.

Now let us solve the surveyor's problem. From

Theorem 4. 6 we see that he should solve

$$\begin{pmatrix} 1 & 0 & 0 & -1 & -1 & 0 \\ 0 & 1 & 0 & 1 & 0 & -1 \\ 0 & 0 & 1 & 0 & 1 & 1 \end{pmatrix} \begin{pmatrix} 1 & 0 & 0 \\ 0 & 1 & 0 \\ 0 & 0 & 1 \\ -1 & 1 & 0 \\ -1 & 0 & 1 \\ 0 & -1 & 1 \end{pmatrix} \begin{pmatrix} A \\ B \\ C \end{pmatrix} = \begin{pmatrix} 1 & 0 & 0 & -1 & -1 & 0 \\ 0 & 1 & 0 & 1 & 0 & -1 \\ 0 & 0 & 1 & 0 & 1 & 1 \end{pmatrix} \begin{pmatrix} 1236 \\ 1941 \\ 2417 \\ 711 \\ 1177 \\ 475 \end{pmatrix}$$

which reduces to

$$\begin{pmatrix} 3 & -1 & -1 \\ -1 & 3 & -1 \\ -1 & -1 & 3 \end{pmatrix} \begin{pmatrix} A \\ B \\ C \end{pmatrix} = \begin{pmatrix} -652 \\ 2177 \\ 4069 \end{pmatrix}.$$

Solving this system we obtain

$$A = 1235.50$$

$$B = 1942.75$$

$$C = 2415.75 .$$

Thus the surveyor reports the heights $A = 1236$ ft., $B = 1943$ ft., $C = 2416$ ft.

EXERCISES.   CHAPTER 5

1.   Find the inverse of the matrix

$$\begin{pmatrix} 3 & -1 & -1 \\ -1 & 3 & -1 \\ -1 & -1 & 3 \end{pmatrix}.$$

2.   Let A and B be $n \times n$ matrices, both positive definite, i.e., $x^* A x > 0$, $x^* B x > 0$ if $x \neq 0$, but A and B are not necessarily symmetric.

(i)   Show that $A + B$ is positive definite.

(ii)   Show that AB need not be positive definite.

Hint: look at 2×2 matrices of the form

$$\begin{pmatrix} \alpha & \beta \\ -\beta & \alpha \end{pmatrix} .$$

(iii) Is $A^2 = AA$ positive definite?   How about when

A is symmetric?

3.   The vector $x^* \in R^n$ which minimizes $(Ax - b)'W(Ax - b)$

is called the least squares solution of $Ax = b$ with

weighting matrix W.   Here W is an $m \times m$ positive

definite symmetric matrix.   Obtain an explicit formula

for x^* in terms of A, W and b using a process of

completing the square as above.

4.   Let us suppose the surveyor in the problem of the three

mountains regards the three measurements made at sea

level as being somewhat more reliable than those made

at the summits of A and B and the two measurements

made at the summit of A as being somewhat more reli-

able than the one made at the summit of B.   He decides

therefore to minimize the quantity

$$4(A - 1236)^2 + 4(B - 1941)^2 + 4(C - 2417)^2$$

$$+ 2(B - A - 711)^2 + 2(C - A - 1177)^2 + (C - B - 475)^2.$$

Use the result of problem 3 to determine which values he should now report for A, B and C.

5. A certain function $f : R^1 \to R^1$ is known to have values $f(x_i) = f_i$, $i = 1, 2, \ldots, m$. It is desired to approximate f with a <u>least squares linear fit</u>, i. e. , to find a linear function $p(x) = ax + b$ such that

$$\sum_{i=1}^{m} (f(x_i) - p(x_i))^2$$

is as small as possible. Show how the coefficients a and b are computed in order to minimize this expression.

6. A buoy has broken loose from its moorings outside Port Bilgewater. It was sighted on Monday, Tuesday, Wednesday and Thursday at the following positions, given relative to P. B.

Monday: 9 mi. south, 11 mi. east.

Tuesday: 17 mi. south, 20 mi. east.

Wednesday: 27 mi. south, 31 mi. east.

Thursday: 36 mi. south, 41 mi. east.

All these measurements are subject to some error. Where would be a reasonable place to look for the buoy on Friday, assuming wind and current remain steady?

7. Determine a polynomial

$$p(x) = a_0 x^4 + a_1 x^3 + a_2 x^2 + a_3 x + a_4$$

of fourth degree providing the best least squares fit to $\sin x$ at the points $x_k = \dfrac{k\pi}{6}$, $k = 0, 1, 2, 3, 4, 5, 6$. That is, determine a_0, a_1, a_2, a_3, a_4 minimizing

$$\sum_{k=0}^{6} (\sin(x_k) - p(x_k))^2.$$

Because of the large amount of computation involved, it is suggested that this be done as a computer programming project.

CHAPTER 5. DIFFERENTIATION AND NEWTON'S METHOD

As we know from elementary calculus, a differentiable function $f(r)$ achieves a maximum or minimum value in an interval $[a, b]$ at a point r^* lying in the interior of that interval only if the derivative of f vanishes there, i. e. , $f'(r^*) = 0$. The equation $f'(r^*) = 0$ can then serve in many cases to identify r^* as in the case of the gasoline mileage problem studied in the Introduction. Our purpose in this chapter is to develop a similar theory to apply to functions of many variables. In order to do this we need a theory of differentiation for such functions. We will assume that the reader is already familiar with the notion of partial derivatives of such functions and with the "chain rule".

141

DEFINITION 5.1. Let $g : R^n \rightarrow R^1$ be defined in some neighborhood of a point $x_0 \in R^n$ with $g(x) > 0$ if $x \neq x_0$. Let $f : R^n \rightarrow R^m$ be defined in a domain D which includes x_0. When we write

$$f(x) = \mathcal{O}(g(x)), \quad x \rightarrow x_0$$

we mean that the quotient $\dfrac{\|f(x)\|}{g(x)}$ is bounded near x_0 in D, i. e. there are numbers K and δ, both greater than zero, such that

$$\frac{\|f(x)\|}{g(x)} \leq K, \quad x \in D, \quad 0 < \|x - x_0\| < \delta.$$

When we write

$$f(x) = o(g(x)), \quad x \rightarrow x_0$$

we mean that $\displaystyle\lim_{\substack{x \rightarrow x_0 \\ x \in D}} \frac{\|f(x)\|}{g(x)} = 0$.

The "order symbols" $\mathcal{O}$ and o will be very convenient. The notations

$$f(x) = h(x) + \mathcal{O}(g(x)), \quad x \rightarrow x_0,$$

$$f(x) = h(x) + o(g(x)), \quad x \rightarrow x_0,$$

can be used to indicate that $f(x) = h(x)$ plus a remainder

term which remains bounded relative to $g(x)$, or tends to

zero faster than $g(x)$, respectively, as x tends to x_0 in

the domain of definition of $f(x)$. For example,

$$e^x = 1 + \mathcal{O}(x), \ x \to 0, \qquad \cos x = 1 + \mathit{o}(x), \ x \to 0.$$

If $f : R^2 \to R^2$ is given by

$$f(x, y) = \left(\frac{\sin \frac{\pi}{2}x + \cos \frac{\pi}{2} y}{\sqrt{1 + (x-1)^2 + (y-1)^3}} \right).$$

then

$$f(x, y) = \begin{pmatrix} 1 - \frac{\pi}{2}(y-1) \\ \\ 1 \end{pmatrix} + \mathit{o}\left((x-1)^2 + (y-1)^2 \right), \ (x, y) \mapsto (1,1).$$

We need the following basic result from vector space

theory.

LEMMA 5.1. (The Schwartz Inequality). If x and y are

two vectors in R^n then

$$|x'y| \le \|x\|_e \|y\|_e$$

and equality holds if and only if there are real numbers α

and β, not both zero, such that

$$\alpha x + \beta y = 0.$$

<u>Proof.</u> If either $x = 0$ or $y = 0$ the desired inequality is obvious. We assume, then, that $x \neq 0$, $y \neq 0$. For real α and β

$$0 \leq \|\alpha x + \beta y\|_e^2 = (\alpha x + \beta y)' (\alpha x + \beta y)$$

$$= \alpha^2 x'x + 2\alpha\beta x'y + \beta^2 y'y.$$

Setting $\alpha = \|y\|_e$, $\beta = -\|x\|_e$, this becomes

$$0 \leq \|y\|_e^2 \|x\|_e^2 - 2\|y\|_e \|x\|_e x'y + \|x\|_e^2 \|y\|_e^2.$$

Since $x \neq 0$ and $y \neq 0$, $\|x\|_e \|y\|_e \neq 0$ and we can divide by $\|x\|_e \|y\|_e$ obtaining

$$0 \leq 2\|x\|_e \|y\|_e - 2x'y$$

and thus

$$x'y \leq \|x\|_e \|y\|_e.$$

Repeating the above computations with $\alpha = \|y\|_e$, $\beta = \|x\|_e$, we obtain

$$-x'y \leq \|x\|_e \|y\|_e$$

and combining out two inequalities we have

$$|x'y| \leq \|x\|_e \|y\|_e.$$

Now $|x'y| = \|x\|_e \|y\|_e$ if and only if $x'y = \|x\|_e \|y\|_e$ or $-x'y = \|x\|_e \|y\|_e$ and, working backwards, we see that these equations can hold if and only if $\|y\|_e x - \|x\|_e y = 0$ or $\|y\|_e x + \|x\|_e y = 0$, respectively. This completes the proof.

DEFINITION 5.2. Let $f : R^n \to R^m$, i.e. for $x \in R^n$,

$$f(x) = \begin{pmatrix} f^1(x) \\ f^2(x) \\ \vdots \\ f^m(x) \end{pmatrix} = \begin{pmatrix} f^1(x^1, x^2, \ldots, x^n) \\ f^2(x^1, x^2, \ldots, x^n) \\ \vdots \\ f^m(x^1, x^2, \ldots, x^n) \end{pmatrix}.$$

(To save space we have written $f^j(x) = f^j(x^1, x^2, \ldots, x^n)$ rather than $f^j \begin{pmatrix} x^1 \\ x^2 \\ \vdots \\ x^n \end{pmatrix}$.) Assume that all of the functions

$f^j(x)$ possess partial derivatives $\dfrac{\partial f^j}{\partial x^1}(x_0)$, $\dfrac{\partial f^j}{\partial x^2}(x_0), \ldots,$

$\dfrac{\partial f^j}{\partial x^n}(x_0)$ at the point $x_0 = \begin{pmatrix} x_0^1 \\ x_0^2 \\ \vdots \\ x_0^n \end{pmatrix}$, $j = 1, 2$. We define the

$m \times n$ matrix $\dfrac{\partial f}{\partial x}(x_0)$ by

$$\frac{\partial f}{\partial x}(x_0) = \begin{pmatrix} \dfrac{\partial f^1}{\partial x^1}(x_0) & \dfrac{\partial f^1}{\partial x^2}(x_0) & \cdots & \dfrac{\partial f^1}{\partial x^n}(x_0) \\[2ex] \dfrac{\partial f^2}{\partial x^1}(x_0) & \dfrac{\partial f^2}{\partial x^2}(x_0) & \cdots & \dfrac{\partial f^2}{\partial x^n}(x_0) \\[2ex] \vdots & \vdots & & \vdots \\[2ex] \dfrac{\partial f^m}{\partial x^1}(x_0) & \dfrac{\partial f^m}{\partial x^2}(x_0) & \cdots & \dfrac{\partial f^m}{\partial x^n}(x_0) \end{pmatrix}.$$

When $m = 1$, $\dfrac{\partial f}{\partial x}(x_0)$ becomes a row vector

$$\frac{\partial f}{\partial x}(x_0) = (\frac{\partial f}{\partial x^1}(x_0), \frac{\partial f}{\partial x^2}(x_0), \ldots, \frac{\partial f}{\partial x^n}(x_0))$$

called the gradient of the scalar valued function $f = f^1$ at

the point x_0. When $n = 1$ we have a column vector

$$\frac{df}{dx}(x_0) = \begin{pmatrix} \frac{df^1}{dx}(x_0) \\ \frac{df^2}{dx}(x_0) \\ \vdots \\ \frac{df^m}{dx}(x_0) \end{pmatrix}.$$

When both $m > 1$ and $n > 1$ the matrix $\frac{\partial f}{\partial x}(x_0)$ is called the Jacobian of f at x_0.

EXAMPLES.

(i) $f : R^2 \to R^2$ is defined by $f^1(x, y) = \sin \frac{\pi}{2}x + \cos \frac{\pi}{2}y$, $f^2(x, y) = \sqrt{1 + (x-1)^2 + (y-1)^2}$. Then

$$\frac{\partial f^1}{\partial x}(0, 0) = \frac{\pi}{2}\cos(\frac{\pi}{2}, 0) = \frac{\pi}{2}.$$

$$\frac{\partial f^1}{\partial y}(0,0) = -\frac{\pi}{2}\sin(\frac{\pi}{2}, 0) = 0$$

$$\frac{\partial f^2}{\partial x}(0,0) = \frac{1}{2}\frac{2(0-1}{\sqrt{1+(0-1)^2+(0-1)^2}} = \frac{-1}{\sqrt{3}}$$

$$\frac{\partial f^2}{\partial y}(0,0) = \frac{1}{2}\frac{2(0-1)}{\sqrt{1+(0-1)^2+(0-1)^2}} = \frac{-1}{\sqrt{3}}$$

and thus the Jacobian of f is $\begin{pmatrix} \dfrac{\pi}{2} & 0 \\ -\dfrac{1}{\sqrt{3}} & -\dfrac{1}{\sqrt{3}} \end{pmatrix}$.

(ii)   Let $f : R^n \to R^m$ be given by $f(x) = Ax + b$ where A is an $m \times n$ matrix and b is an m-vector. Then

$$f^j(x) = a_1^j x^1 + a_2^j x^2 + \ldots + a_n^j x^n + b^j$$

so that

$$\frac{\partial f^j}{\partial x^i}(x) = a_i^j \quad \text{for all} \quad x \in R^n$$

and thus $\dfrac{\partial f}{\partial x}(x) = A$ for all $x \in R^n$.

(iii) Let $f : R^n \to R^1$ be given by

$$f(x) = x'Ax + b'x + c$$

where A is an $n \times n$ matrix, not necessarily symmetric, $b \in R^n$ and $c \in R^1$. Then

$$f(x) = \sum_{j=1}^{n} a_j^j (x^j)^2 + \sum_{\substack{j=1 \\ i \ne j}}^{n} (\sum_{i=1}^{n} a_i^j x^j x^i) + \sum_{j=1}^{n} b^j x^j + c$$

and hence

$$\frac{\partial f}{\partial x^k}(x) = 2a_k^k x^k + \sum_{\substack{i=1 \\ i \neq k}}^{n} a_i^k x^i + \sum_{\substack{j=1 \\ j \neq k}}^{n} a_k^j x^j + b^k$$

$$= \sum_{i=1}^{n} a_i^k x^i + \sum_{j=1}^{n} a_k^j x^j + b^k .$$

From these it is clear that, for all $x \in R^n$,

$$\frac{\partial f}{\partial x}(x) = (\frac{\partial f}{\partial x^1}(x), \frac{\partial f}{\partial x^2}(x), \ldots, \frac{\partial f}{\partial x^n}(x)) = x'(A' + A) + b' .$$

If A is symmetric, then

$$\frac{\partial f}{\partial x}(x) = 2x'A + b' .$$

THEOREM 5.1. Let $f : R^n \to R^m$ have continuous partial

derivatives $\dfrac{\partial f^j}{\partial x^i}(x)$, $j = 1, 2, \ldots, m$, $i = 1, 2, \ldots, n$ for x in

some open set containing $x_0 \in R^n$. Then

$$f(x) = f(x_0) + \frac{\partial f}{\partial x}(x_0)(x - x_0) + \theta(\|x - x_0\|), \quad x \to x_0 .$$

Proof. Let N be a spherical neighborhood of x_0 contained

in the open set mentioned in the theorem. If we take $x \in N$

then the straight line segment joining x_0 to x in R^n also

lies in N. Points on this line segment have the representation

$$x(\lambda) = \lambda x + (1 - \lambda)x_0, \qquad 0 \le \lambda \le 1.$$

That is, $x(\lambda)$ is a function from R^1 into R^n whose values make up the straight line segment in question. We have $x(0) = x_0$, $x(1) = x$, $x(\frac{1}{2}) = \frac{1}{2}x + \frac{1}{2}x_0$. Then $f^j(x(\lambda))$ is a scalar valued function of the scalar variable λ for $0 \le \lambda \le 1$ which we will call $g^j(\lambda)$. Thus $g^j(0) = f^j(x_0)$, $g^j(1) = f^j(x)$. According to the mean value theorem

$$g^j(1) - g^j(0) = \frac{dg^j}{d\lambda}(\lambda_j)(1 - 0) = \frac{dg^j}{d\lambda}(\lambda_j), \qquad 0 < \lambda_j < 1.$$

We use the chain rule to compute the derivative

$$\frac{dg^j}{d\lambda}(\lambda_j) = \frac{d}{d\lambda}(f^j(x(\lambda)))\Big|_{\lambda = \lambda_j} = \Big(\sum_{i=1}^{n}\frac{\partial f^j}{\partial x^i}(x(\lambda))\frac{dx^i}{d\lambda}(\lambda))\Big)\Big|_{\lambda = \lambda_j}$$

$$= \sum_{i=1}^{n}\frac{\partial f^j}{\partial x^i}(x(\lambda_j))(x^i - x_0^i).$$

Thus

$$f^j(x) - f^j(x_0) = \sum_{i=1}^{n}\frac{\partial f^j}{\partial x^i}(x(\lambda_j))(x^i - x_0^i)$$

$$= \sum_{i=1}^{n}\frac{\partial f^j}{\partial x^i}(x_0)(x^i - x_0^i) + \sum_{i=1}^{n}(\frac{\partial f^j}{\partial x^i}(x(\lambda_j)) - \frac{\partial f^j}{\partial x^i}(x_0))(x^i - x_0^i).$$

Now $x(\lambda_j)$ lies on the line segment between x and x_0 and hence is at least as close to x_0 as x is. Since $\frac{\partial f^j}{\partial x^i}(x)$ is

a continuous function of x near x_0 we can make the dif-

ference $\frac{\partial f^j}{\partial x^i}(x(\lambda_j)) - \frac{\partial f^j}{\partial x^i}(x_0)$ as small as we wish by choosing

x sufficiently close to x_0. Now the sum $\sum_{i=1}^{n}(\frac{\partial f^j}{\partial x^i}(x(\lambda_j)) -$

$\frac{\partial f^j}{\partial x^i}(x_0))(x^i-x_0^i)$ is the inner product of $x-x_0$ with a vector

whose components are $\frac{\partial f^j}{\partial x^i}(x(\lambda_j)) - \frac{\partial f^j}{\partial x^i}(x_0)$. Thus, from

Lemma 5.1,

$$\left|\sum_{i=1}^{n}(\frac{\partial f^j}{\partial x^i}(x(\lambda_j))-\frac{\partial f^j}{\partial x^i}(x_0))(x^i-x_0^i)\right|$$

$$\leq [\sum_{i=1}^{n}(\frac{\partial f^j}{\partial x^i}(x(\lambda_j))-\frac{\partial f^j}{\partial x^i}(x_0))^2]^{\frac{1}{2}}\|x-x_0\|_e$$

and we see that, for $x \to x_0$,

$$\sum_{i=1}^{n}(\frac{\partial f^j}{\partial x^i}(x(\lambda_j))-\frac{\partial f^j}{\partial x^i}(x_0))(x^i-x_0^i)= \sigma(\|x-x_0\|_e)= \sigma(\|x-x_0\|),$$

no matter which norm is used. (Note exercise 1, Chapter 1.)

Thus

$$f^j(x)-f^j(x_0)= \sum_{i=1}^{n}\frac{\partial f^j}{\partial x^i}(x_0)(x^i-x_0^i)+ \sigma(\|x-x_0\|), \quad x \to x_0.$$

Since this is true for each $j = 1, 2, \ldots, m$, noting that

$\displaystyle\sum_{i=1}^{n} \frac{\partial f^j}{\partial x^i} (x_0)(x^i - x_0^i)$ is the j-th component of $\frac{\partial f}{\partial x}(x_0)(x-x_0)$,

we have

$$f(x) - f(x_0) = \frac{\partial f}{\partial x}(x_0)(x-x_0) + \boldsymbol{\sigma}(\|x-x_0\|), \quad x \to x_0$$

and this completes the proof.

The function $f(x_0) + \frac{\partial f}{\partial x}(x_0)(x-x_0)$, composed of a constant and a linear function, is called the <u>linear approximation</u> to f at x_0. To give some idea of what this means we consider the case $m = 1$, $n = 2$. The graph of f is then a two-dimensional surface in R^3 as shown in Figure 7.

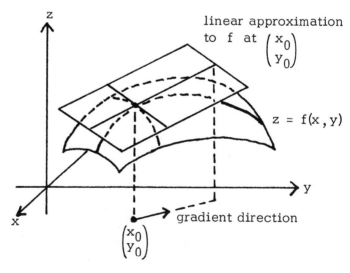

Figure 7

The graph of $f(x_0) + \frac{\partial f}{\partial x}(x_0)(x - x_0)$ is a two dimensional

plane meeting the surface which is the graph of f tangential-

ly at the point $\begin{pmatrix} x_0^1 \\ x_0^2 \\ f(x_0^1, x_0^2) \end{pmatrix}$ in R^3. If we conceive of the

graph of f as an inverted bowl we can think of the graph of

its linear approximation as a plate balanced on the bowl at

this point. For larger values of m and n it is difficult to

draw pictures but the situation is similar.

The next theorem is one of the most basic results in

optimization theory.

THEOREM 5.2. Let $f : R^n \to R^1$ have a local (or global)

minimum in a set S at a point $x^* \in S^o$, the interior of S.

Then, if f is continuously differentiable in some neighbor-

hood of x^*, the gradient $\frac{\partial f}{\partial x}(x^*) = 0$.

Proof. Assume the contrary, that $\frac{\partial f}{\partial x}(x^*) \neq 0$. Let $x = x^* -$

$\lambda \left(\frac{\partial f}{\partial x}(x^*)\right)'$, $\lambda > 0$, which lies near x^* if λ is small. Ac-

cording to Theorem 5.1

$$f(x) = f(x^*) + \frac{\partial f}{\partial x}(x^*)(x - x^*) + \mathscr{O}(\|x - x^*\|)$$

$$= f(x^*) + \frac{\partial f}{\partial x}(x^*)(-\lambda(\frac{\partial f}{\partial x}(x^*))') + \mathscr{O}(\|-\lambda(\frac{\partial f}{\partial x}(x^*))'\|)$$

$$= f(x^*) - \lambda \|(\frac{\partial f}{\partial x}(x^*))'\|_e^2 + \mathscr{O}(\lambda)$$

$$= f(x^*) - \lambda[\|(\frac{\partial f}{\partial x}(x^*))'\|_e^2 - \frac{\mathscr{O}(\lambda)}{\lambda}] .$$

By definition, $\frac{\mathscr{O}(\lambda)}{\lambda}$ approaches zero as $\lambda \to 0$. On the

other hand, since $\frac{\partial f}{\partial x}(x^*)$ has been assumed different from

the zero vector, $\|(\frac{\partial f}{\partial x}(x^*))'\|_e^2 > 0$. We conclude then that

$$f(x) = f(x^* - \lambda(\frac{\partial f}{\partial x}(x^*))') < f(x^*),$$

for sufficiently small positive values of λ, contradicting our

assumption that x^* was a local minimum of f in S. We

use the assumption that $x^* \in S^o$ to guarantee that

$x^* - \lambda(\frac{\partial f}{\partial x}(x^*))' \in S$ if λ is small. This contradiction shows

that $\frac{\partial f}{\partial x}(x^*)$ must be 0 and the proof is complete.

The vector $(\frac{\partial f}{\partial x}(x_0))'$, when different from zero, indi-

cates the direction of most rapid increase of the function f

and $-(\frac{\partial f}{\partial x}(x_0))'$ indicates the direction of most rapid decrease.

Given any unit vector $\eta \in R^n$, "unit vector" meaning $\|\eta\|_e = $

1, we define the <u>directional derivative of f in the direction</u>

of η by

$$\frac{df}{d\eta}(x_0) = \frac{\partial f}{\partial x}(x_0)\eta.$$

The real number $\frac{df}{d\eta}(x_0)$ indicates the rate of increase, or decrease, of f as we leave x_0 along a ray pointing in the direction of the vector η. According to the Schwartz inquality, Lemma 5.1

$$\left|\frac{df}{d\eta}(x_0)\right| = \left|\frac{\partial f}{\partial x}(x_0)\eta\right| = \left|((\frac{\partial f}{\partial x}(x_0))')'\eta\right|$$

$$\leq \left\|(\frac{\partial f}{\partial x}(x_0))'\right\|_e \|\eta\|_e \leq \left\|(\frac{\partial f}{\partial x}(x_0))'\right\|_e,$$

and equality holds only if there exist real α, β such that

$$\alpha\eta + \beta(\frac{\partial f}{\partial x}(x_0))' = 0,$$

i.e. $\quad \eta = \frac{-\beta}{\alpha}(\frac{\partial f}{\partial x}(x_0))'.$

Thus the absolute value of $\frac{df}{d\eta}(x_0)$ is maximized when η has the same, or the opposite, direction as the vector $(\frac{\partial f}{\partial x}(x_0))'.$

EXAMPLE. Let $f(x,y) = x^2 + 2xy - 3y^2$. Then the gradient is the row vector $(2x + 2y, 2x - 6y)$. At $(x,y) = (1,1)$ this becomes $(4, -4)$. Given $0 \leq \theta < 2\pi$, the unit vector

$\eta = \begin{pmatrix} \cos \theta \\ \sin \theta \end{pmatrix}$ makes an angle θ with the positive x axis.

The directional derivative of $x^2 + 2xy - 3y^2$ in the direction

of η is $(4, -4) \begin{pmatrix} \cos \theta \\ \sin \theta \end{pmatrix} = 4(\cos \theta - \sin \theta)$. To determine

when this directional derivative has its maxima and minima

we differentiate

$$\frac{d}{d\theta}(\cos \theta - \sin \theta) = -\sin \theta - \cos \theta$$

which vanishes when $\sin \theta = -\cos \theta$, i.e. when $\theta = \dfrac{7\pi}{4}$ or

$\dfrac{3\pi}{4}$. Thus the directional derivative has maximum absolute

value when $\eta = \begin{pmatrix} \dfrac{1}{\sqrt{2}} \\ -\dfrac{1}{\sqrt{2}} \end{pmatrix}$ or $\eta = \begin{pmatrix} -\dfrac{1}{\sqrt{2}} \\ \dfrac{1}{\sqrt{2}} \end{pmatrix}$, both of which are

multiples of $(4, -4)' = \begin{pmatrix} 4 \\ -4 \end{pmatrix}$. The largest positive value

occurs for $\eta = \begin{pmatrix} \dfrac{1}{\sqrt{2}} \\ -\dfrac{1}{\sqrt{2}} \end{pmatrix} = \dfrac{1}{4\sqrt{2}} \begin{pmatrix} 4 \\ -4 \end{pmatrix}$. Thus we see in this

example that the direction of most rapid increase of $f(x, y)$

is the direction of the transposed gradient vector.

  Theorem 5.2 implies that if f has a local minimum in

S at $x^* \in S^o$ then the directional derivative vanishes at x^*

for every direction η.

COROLLARY TO THEOREM 5. 2. If f satisfies the same hypotheses as in Theorem 5. 2 and has a local (or global) maximum in S at $x^* \in S^o$ then $\frac{\partial f}{\partial x}(x^*) = 0$.

Proof. Apply Theorem 5. 2 to the function $- f$.

Theorem 5. 2 and its Corollary provide <u>necessary</u> conditions which must be fulfilled if f is to have a local minimum or maximum at x^* . The condition $\frac{\partial f}{\partial x}(x^*) = 0$ is not <u>sufficient</u> to guarantee that x^* is either a maximum or a minimum of f . It is only necessary to cite the example $f(x, y) = x^2 - y^2$, whose gradient $(2x, -2y)$ vanishes at $(0, 0)$, which has neither a maximum nor a minimum there. Nevertheless, if used with discretion, the equation $\frac{\partial f}{\partial x}(x^*) = 0$ can be useful in locating maxima and minima.

EXAMPLE. Three men measure a table top. The first measures its length and declares it to be 5. 12 ft. , the second measures its width and finds it to be 4. 06 ft. and the third finds the diagonal measurement to be 6. 43 ft. What length, ℓ , and width, w , should the men agree upon for the table?

The three measurements are described by the equations

$$\ell = 5.12$$

$$w = 4.06$$

$$\sqrt{\ell^2 + w^2} = 6.43,$$

which are slightly inconsistent. The least squares principle

indicates that ℓ and w should be chosen so as to minimize

$$f(\ell, w) = (\ell - 5.12)^2 + (w - 4.06)^2 + (\sqrt{\ell^2 + w^2} - 6.43)^2.$$

To apply Theorem 5.2 we compute

$$\frac{\partial f}{\partial \ell} = 2(\ell - 5.12) + 2(\sqrt{\ell^2 + w^2} - 6.43)\frac{\ell}{\sqrt{\ell^2 + w^2}} = 0$$

$$\frac{\partial f}{\partial w} = 2(w - 4.06) + 2(\sqrt{\ell^2 + w^2} - 6.43)\frac{w}{\sqrt{\ell^2 + w^2}} .$$

Thus we wish to solve

$$2\ell - 5.12 = \frac{6.43\,\ell}{\sqrt{\ell^2 + w^2}}$$

$$2w - 4.06 = \frac{6.43w}{\sqrt{\ell^2 + w^2}} .$$

Squaring both sides of the two equations and adding we have

$$4\ell^2 - 20.48\ell + 26.21 + 4w^2 - 16.24w + 16.48 = (6.43)^2 \frac{\ell^2 + w^2}{\ell^2 + w^2} = 41.34 .$$

On the other hand we can write

$$\ell \left(2 - \frac{6.43}{\sqrt{\ell^2 + w^2}}\right) = 5.12$$

$$w \left(2 - \frac{6.43}{\sqrt{\ell^2 + w^2}}\right) = 4.06$$

so that $\frac{\ell}{w} = \frac{5.12}{4.06} = 1.26$, i. e. $\ell = 1.26w$. Thus

$$4(1.26w)^2 - 20.48(1.26w) + 4w^2 - 16.24w + 1.35 = 0 ,$$

i. e. $10.36w^2 - 42.04w + 1.35 = 0 .$

From the quadratic formula

$$w = \frac{42.04 \pm \sqrt{(42.04)^2 - 4(10.36)(1.35)}}{2(10.36)}$$

$$= 2.03 \pm 2.02 = 4.05 \text{ ft.}$$

Then

$$\ell = 1.26 w = 5.11 \text{ ft.}$$

It should be noted that we were very lucky in the a-
bove example. The equations $\frac{\partial f}{\partial \ell} = 0$, $\frac{\partial f}{\partial w} = 0$ are both nonlinear and special tricks are needed to solve them. In gen-
eral we will not be able to solve these equations by analytical

methods.   Thus we need numerical methods for solving gen-

eral equations of the form

$$g(x) = 0$$

where g is a function $g : R^n \to R^n$.

Let us suppose that $g(x) = \begin{pmatrix} g^1(x^1, x^2, \ldots, x^n) \\ g^2(x^1, x^2, \ldots, x^n) \\ \vdots \\ g^n(x^1, x^2, \ldots, x^n) \end{pmatrix}$ is

continuously differentiable in an open set which includes a

point x^* where $g(x^*) = 0$.   Assuming x_0 lies fairly close

to x^* we can write

$$0 = g(x^*) = g(x_0) + \frac{\partial g}{\partial x}(x_0)(x^* - x_0) + \sigma(\|x^* - x_0\|)$$

or $\frac{\partial g}{\partial x}(x_0)(x^* - x_0) = -g(x_0) + \sigma(\|x^* - x_0\|)$.

Were it not for the fact that the right hand side involves x^*,

we would have an equation of the form

$$A(x^* - x_0) = b$$

where $A = \frac{\partial g}{\partial x}(x_0)$ is an $n \times n$ matrix, and, if A were non-

singular, we could solve for $x^* - x_0$ and thereby obtain x^*, since x_0 is known. This suggests that we attempt to approximate x^* by a vector x_1, where

$$\frac{\partial g}{\partial x}(x_0)(x_1 - x_0) = -g(x_0).$$

If the term $\mathscr{O}(\|x^* - x_0\|)$ were small we might expect that x_1 would lie closer to x^* than x_0 did. Then we could compute another vector x_2 by solving

$$\frac{\partial g}{\partial x}(x_k)(x_{k+1} - x_k) = -g(x_k).$$

Under appropriate assumptions we shall prove that $\lim_{k \to \infty} x_k = x^*$ and thus the above method, which is known as Newton's method, provides a numerical technique for approximating the solution x^* of the equation $g(x) = 0$.

When $n = 1$ Newton's method can be made plausible by the following argument. As indicated in Figure 8 we suppose $g(x)$ to be a real valued function of the real variable x, possessing a simple zero at a point x^*. We assume also that $\frac{dg}{dx}(x^*) \neq 0$.

Being unable to determine x^*, the point where the graph of g crosses the x axis, we replace the curve which is the graph of g by the line which is tangent to g at a certain point x_0. This line is the graph of the linear function

$$h(x) = \frac{dg}{dx}(x_0)(x - x_0) + g(x_0).$$

We let x_1 be the point where this straight line crosses the x axis, i. e.,

$$0 = h(x_1) = \frac{dg}{dx}(x_0)(x_1 - x_0) + g(x_0),$$

i. e.,

$$x_1 = x_0 - \frac{g(x_0)}{\frac{dg}{dx}(x_0)}.$$

Obtaining x_1 in this way we next construct the line tangent to the graph of g at x_1 and we let x_2 be the point where this line crosses the x-axis, etc.

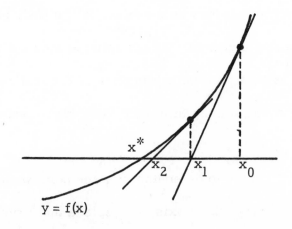

Figure 8

We shall give an example of the use of Newton's method before going on to prove its validity.

EXAMPLE. Solve the equations

$$\sin x + 2y = 1$$

$$2x + \cos y = 2 \quad .$$

Here $g_1(x, y) = \sin x + 2y - 1$, $g_2(x, y) = 2x + \cos y - 2$ while

$$\frac{\partial g_1}{\partial x}(x, y) = \cos x \qquad \frac{\partial g_1}{\partial y}(x, y) = 2$$

$$\frac{\partial g_2}{\partial x}(x, y) = 2 \qquad \frac{\partial g_2}{\partial y}(x, y) = -\sin y \quad .$$

Thus Newton's method is

$$\begin{pmatrix} \cos x_k & 2 \\ 2 & -\sin y_k \end{pmatrix} \begin{pmatrix} y_{k+1} - x_k \\ y_{k+1} - y_k \end{pmatrix} = \begin{pmatrix} 1 - \sin x_k - 2y\,k \\ 2 - 2x_k - \cos y_k \end{pmatrix} \quad .$$

We start with $x = y_0 = .500$, which gives

$$\begin{pmatrix} .877 & 2 \\ 2 & -.479 \end{pmatrix} \begin{pmatrix} x_1 - .500 \\ y_1 - .500 \end{pmatrix} = \begin{pmatrix} -.479 \\ -.123 \end{pmatrix}$$

or

$$.877(x_1 - .500) + 2(y_1 - .500) = -.479$$

$$2(x_1 - .500) - .479(y_1 - .500) = .123 .$$

We can solve this system of two linear equations in two unknowns by the usual elimination technique to obtain

$$x_1 - .500 = .004 , \qquad y_1 - .500 = -.241$$

or

$$x_1 = .504 , \qquad y_1 = .259 .$$

Applying Newton's method again we have

$$\begin{pmatrix} .875 & 2 \\ 2 & -.255 \end{pmatrix} \begin{pmatrix} x_2 - .504 \\ y_2 - .259 \end{pmatrix} = \begin{pmatrix} .000 \\ .027 \end{pmatrix}$$

which yields

$$x_2 - .504 = .013 \qquad y_2 - .259 = -.005$$

or

$$x_2 = .517 , \quad y_2 = .254 .$$

Once more:

$$\begin{pmatrix} .869 & 2 \\ 2 & -.252 \end{pmatrix} \begin{pmatrix} x_3 - .517 \\ y_3 - .254 \end{pmatrix} = \begin{pmatrix} -.001 \\ .001 \end{pmatrix}$$

yields

$$x_3 - .517 = .001, \quad y_3 - .254 = -.001$$

or

$$x_3 = .518, \quad y_3 = .253.$$

The right hand side for the next system of equations is

$$-\begin{pmatrix} g_1(x_3, y_3) \\ g_2(x_3, y_3) \end{pmatrix} = \begin{pmatrix} .000 \\ .000 \end{pmatrix}$$ and we therefore stop with the values

x_3, y_3 just obtained.

We need some preliminary material before we can prove

the validity of Newton's method.

DEFINITION 5.3.   Let A be an $m \times n$ matrix.   We define

the _norm_ of A, $\|A\|$, by

$$\|A\| = \max_{\|x\| \neq 0} \left\{ \frac{\|Ax\|}{\|x\|} \right\}.$$

It is clear then that for all $x \in R^n$, $\|Ax\| \leq \|A\| \|x\|$.

Depending on whether we use $\| \ \|_e$, $\| \ \|_s$ or $\| \ \|_t$ for

x and Ax we obtain $\|A\|_e$, $\|A\|_s$ or $\|A\|_t$, respectively.

EXAMPLE.   Let A be an $m \times n$ matrix with entries a_j^i,

$i = 1, 2, \ldots, m$, $j = 1, 2, \ldots, n$. Let $Ax = y = \begin{pmatrix} y^1 \\ \vdots \\ y^m \end{pmatrix}$. Then

$$\|Ax\|_e^2 = \sum_{i=1}^{m} (y^i)^2 = \sum_{i=1}^{m} (\sum_{j=1}^{n} a_j^i x^j)^2 .$$

Using Lemma 5.1, the Schwartz inequality, and regarding $\sum_{j=1}^{n} a_j^i x^j$ as the inner product of two vectors we have

$$\sum_{i=1}^{m} (\sum_{j=1}^{n} a_j^i x^j)^2 \leq \sum_{i=1}^{m} ((\sum_{j=1}^{n} (a_j^i)^2)(\sum_{j=1}^{n} (x^j)^2))$$

$$= (\sum_{\substack{i=1 \\ j=1}}^{m,n} (a_j^i)^2 \|x\|_e^2$$

and we conclude

$$\|Ax\|_e \leq (\sum_{\substack{i=1 \\ j=1}}^{m,n} (a_j^i)^2)^{1/2} \|x\|_e, \qquad x \in R^n.$$

It follows that the Euclidean norm of a matrix is less than or equal to the square root of the sum of the squares of the entries of A. This is a very generous estimate in most cases, e.g., if $A = I$ is the $n \times n$ identity then $\|I\|_e = 1$ but what we have done here shows only that $\|I\|_e \leq \sqrt{n}$. It is difficult to compute $\|A\|_e$ exactly. See the exercises at the end of this chapter for comparable results concerning $\|A\|_s$ and $\|A\|_t$.

Whatever norm is used, one easily verifies

$$\|\lambda A\| = |\lambda| \, \|A\|, \quad \lambda \text{ real,}$$

$$\|A + B\| \leq \|A\| + \|B\|,$$

$$\|AB\| \leq \|A\| \, \|B\|,$$

provided the dimension of A, B are such that $A + B$ or AB are defined.

We say that a sequence $\{A_k\}$ of $m \times n$ matrices converges to the $m \times n$ matrix A if $\lim\limits_{k \to \infty} \|A - A_k\| = 0$. A matrix valued function $A(\xi)$ is continuous at a point ξ_0 if

$$\lim\limits_{\|\xi - \xi_0\| \to 0} \|A(\xi) - A(\xi_0)\| = 0.$$

LEMMA 5.2. Let $A(\xi)$ be an $n \times n$ matrix valued function of the vector $\xi \in R^r$ which is continuous at ξ_0. If $A(\xi_0)$ is non-singular, so that $A(\xi_0)^{-1}$ exists, then $A(\xi)^{-1}$ exists for ξ sufficiently near ξ_0 and $A(\xi)^{-1}$ is continuous at ξ_0.

Proof. We write

$$A(\xi) = A(\xi_0) + A(\xi) - A(\xi_0) = A(\xi_0)(I + K(\xi))$$

where $K(\xi) = A(\xi_0)^{-1}(A(\xi) - A(\xi_0))$. Then

$$\|K(\xi)\| \le \|A(\xi_0)^{-1}\| \, \|A(\xi) - A(\xi_0)\|$$

and $\lim\limits_{\|\xi - \xi_0\| \to 0} \|K(\xi)\| = 0$.   Consider the infinite series of matrices

$$I - K(\xi) + (K(\xi))^2 - (K(\xi))^3 + \ldots + (-1)^k (K(\xi))^k + \ldots \; .$$

It is easy to verify that if $\|K(\xi)\| < 1$ this series converges to an $n \times n$ matrix $H(\xi)$. We claim $H(\xi) = (I + K(\xi))^{-1}$. For

$$(I + K(\xi))H(\xi) = (I + K(\xi)) \sum_{k=0}^{\infty} (-1)^k (K(\xi))^k \, ,$$

where we take $(K(\xi))^0 = I$. But

$$(I + K(\xi)) \sum_{k=0}^{\infty} (-1)^k (K(\xi))^k = I + \sum_{k=1}^{\infty} (-1)^k (K(\xi))^k$$

$$+ \sum_{k=0}^{\infty} (-1)^k (K(\xi))^{k+1} = I + \sum_{k=1}^{\infty} (-1)^k ((K(\xi))^k - (K(\xi))^k) = I \; .$$

Thus, if we choose ξ close enough to ξ_0 so that $\|A(\xi) -$

$A(\xi_0)\| \le \dfrac{\rho}{\|A(\xi_0)^{-1}\|}$, where $0 < \rho < 1$, then $\|K(\xi)\| \le \rho < 1$

and the series $\displaystyle\sum_{k=0}^{\infty} (-1)^k (K(\xi))^k$ converges to $H(\xi) = (I + K(\xi))^{-1}$.

Then

$$A(\xi)^{-1} = (A(\xi_0)(I + K(\xi))) = H(\xi)A(\xi_0)^{-1},$$

and thus $A(\xi)^{-1}$ exists for ξ close to ξ_0. Also

$$\|A(\xi)^{-1} - A(\xi_0)^{-1}\| = \|(H(\xi) - I)A(\xi_0)^{-1}\|$$

$$\leq \|H(\xi) - I\| \, \|A(\xi_0)^{-1}\|$$

and

$$\|H(\xi) - I\| = \left\|\sum_{k=1}^{\infty} (-1)^k (K(\xi))^k\right\| \leq \sum_{k=1}^{\infty} \|K(\xi)\|^k$$

$$= \frac{\|K(\xi)\|}{1 - \|K(\xi)\|} \quad \text{when} \quad \|K(\xi)\| < 1.$$

Since $\displaystyle\lim_{\|\xi - \xi_0\| \to 0} \frac{\|K(\xi)\|}{1 - \|K(\xi)\|} = 0$ we conclude

$$\lim_{\|\xi - \xi_0\|} \|A(\xi)^{-1} - A(\xi_0)^{-1}\| = 0 \quad \text{and the proof is complete.}$$

THEOREM 5.3. Let $g : R^n \to R^n$ be continuously differentiable

(i.e., $\frac{\partial g}{\partial x}(x)$ is a continuous function of x) in an open set

which includes a point x^* where $g(x^*) = 0$. If $\frac{\partial g}{\partial x}(x^*)$ is is a non-singular matrix then Newton's method

$$\frac{\partial g}{\partial x}(x_k)(x_{k+1} - x_k) = -g(x_k)$$

yields a sequence $\{x_k\}$ converging to x^* as $k \to \infty$, provided x_0 is chosen sufficiently close to x^*.

<u>Proof.</u> Newton's method is described by the formula $x_{k+1} = h(x_k)$ where

$$h(x) = x - (\frac{\partial g}{\partial x}(x))^{-1} g(x),$$

assuming $(\frac{\partial g}{\partial x}(x))^{-1}$ exists. We note that $g(x) = 0$ if and only if $h(x) = x$.

Let x be a point in R^n lying close to x^*. We compute

$$h(x) - h(x^*) = x - x^* - [\frac{\partial g}{\partial x}(x)^{-1} g(x)].$$

The matrix $\frac{\partial g}{\partial x}(x^*)$ is non-singular and $\frac{\partial g}{\partial x}(x)$ is a continuous $n \times n$ matrix valued function of x for x near x^*. Using Lemma 5.2 we see that $\frac{\partial g}{\partial x}(x)$ is non-singular for x near x^* and

$$\frac{\partial g}{\partial x}(x)^{-1} = \frac{\partial g}{\partial x}(x^*)^{-1} + D(x)$$

where $\lim_{\|x-x^*\| \to 0} \|D(x)\| = 0$. Since $g(x^*) = 0$ we have, using the mean value theorem as in the proof of Theorem 5.1, for $j = 1, 2, \ldots, n$,

$$g^j(x) = g^j(x^*) + \frac{\partial g^j}{\partial x}(x(\lambda_j))(x - x^*) = \frac{\partial g^j}{\partial x}(x(\lambda_j))(x - x^*).$$

Since $\frac{\partial g^j}{\partial x}(x)$ is a continuous function of x near x^* we have

$$\frac{\partial g^j}{\partial x}(x(\lambda_j)) = \frac{\partial g^j}{\partial x}(x^*) + e^j(x(\lambda_j))$$

where $\lim_{\|x-x^*\| \to 0} \|e^j(x)\| = 0$. The point $x(\lambda_j)$ lies on the line segment joining x^* to x and thus lies closer to x^* then does x. Therefore $\|e^j(x(\lambda_j))\|$ becomes small as x nears x_0. We let $E(x)$ be the matrix with rows $e^j(x(\lambda))$, $j = 1, 2, \ldots, n$. Then $\lim_{\|x-x^*\| \to 0} \|E(x)\| = 0$ and

$$g(x) = \frac{\partial g}{\partial x}(x^*)(x - x^*) + E(x)(x - x^*).$$

Then, returning to $h(x) - h(x^*)$ we have

$$h(x)-h(x^*) = x-x^* - [(\frac{\partial g}{\partial x}(x^*)^{-1}+D(x))(\frac{\partial g}{\partial x}(x^*)(x-x^*)$$

$$+ E(x)(x-x^*))] = x-x^* - \frac{\partial g}{\partial x}(x^*)^{-1}\frac{\partial g}{\partial x}(x^*)(x-x^*)$$

$$+\frac{\partial g}{\partial x}(x^*)^{-1}E(x)(x-x^*)+D(x)\frac{\partial g}{\partial x}(x^*)(x-x^*)$$

$$=\frac{\partial g}{\partial x}(x^*)^{-1}E(x)(x-x^*)+D(x)\frac{\partial g}{\partial x}(x^*)(x-x^*)$$

since $\frac{\partial g}{\partial x}(x^*)^{-1}\frac{\partial g}{\partial x}(x^*) = I$, the $n \times n$ identity matrix.

Since $\lim\limits_{\|x-x_0\|\to 0} \|E(x)\| = 0$ and $\lim\limits_{\|x-x_0\|\to 0} \|D(x)\| = 0$

we see that, given any ρ, $0 < \rho < 1$, if we take x sufficiently close to x^* then

$$\|\frac{\partial g}{\partial x}(x^*)^{-1}E(x)\| \le \frac{\rho}{2}, \quad \|D(x)\frac{\partial g}{\partial x}(x^*)\| \le \frac{\rho}{2}$$

and, recalling that $h(x) = x^*$, we obtain the estimate

$$\|h(x) - x^*\| = \|h(x) - h(x^*)\| \le \rho\|x-x^*\|$$

so that $h(x)$ lies closer to x^* than does x if x is sufficiently close to x^*.

Now the convergence is easy to prove. We choose x_0 close enough to x^* so that the above estimates apply.

Newton's method gives $x_1 = h(x_0)$ so we have

$$\|x_1 - x^*\| = \|h(x_0) - x^*\| \le \rho \|x_0 - x^*\|.$$

Then x_1 is even closer to x^* so the estimates still hold and, noting that $x_2 = h(x_1)$ we have

$$\|x_2 - x^*\| \le \rho \|x_1 - x^*\| \le \rho^2 \|x_0 - x^*\|.$$

Continuing in this manner we find that for all k

$$\|x_{k+1} - x^*\| \le \rho \|x_k - x^*\|,$$

$$\|x_k - x^*\| \le \rho \|x_0 - x^*\|.$$

Since $\|x_0 - x^*\|$ is a fixed number and $0 < \rho < 1$ we conclude that

$$0 \le \lim_{k \to \infty} \|x_k - x^*\| \le (\lim_{k \to \infty} \rho_k) \|x_0 - x^*\| = 0$$

and hence the sequence $\{x_k\}$ converges to the solution x^* of $g(x^*) = 0$ and the proof is complete.

The condition that $g(x)$ be continuously differentiable near x^* cannot be dispensed with. The real valued function $g(r) = r^{1/3}$ has the property $g(0) = 0$ but Newton's method gives

$$r_{k+1} = r_k - \frac{g(r_k)}{\frac{dg}{dr}(r_k)} = r_k - \frac{r_k^{1/3}}{\frac{1}{3}r_k^{-2/3}} = -2r_k$$

so that r_{k+1} always lies twice as far away from 0 as does r_k and we cannot have $\lim_{k \to \infty} r_k = 0$ no matter how close r_0 lies to 0.

Now let us apply Newton's method to optimization problems. We are given a function $f : R^n \to R^1$ which has a minimum at a point x^* in the interior of the set S. Thus, from Theorem 5.2, $\frac{\partial f}{\partial x}(x^*) = 0$. Since this equation involves row vectors we transpose it to obtain

$$\frac{\partial f}{\partial x}(x^*)' = 0.$$

Thus we wish to solve $g(x) = 0$ where $g : R^n \to R^n$ is given by $g(x) = \frac{\partial f}{\partial x}(x^*)'$. Newton's method

$$\frac{\partial g}{\partial x}(x_k)(x_{k+1} - x_k) = -g(x_k)$$

becomes

$$\frac{\partial}{\partial x}(\frac{\partial f}{\partial x}(x_k)')(x_{k+1} - x_k) = -\frac{\partial f}{\partial x}(x_k).$$

Now if $f(x) = f(x^1, x^2, \ldots, x^n)$, $\dfrac{\partial f'}{\partial x} = \begin{pmatrix} \dfrac{\partial f}{\partial x^1} \\ \vdots \\ \dfrac{\partial f}{\partial x^n} \end{pmatrix}$ and $\dfrac{\partial}{\partial x}\left(\dfrac{\partial f'}{\partial x}\right)$

is the matrix $\dfrac{\partial^2 f}{\partial x^2}$ defined by

$$\frac{\partial^2 f}{\partial x^2} = \begin{pmatrix} \dfrac{\partial^2 f}{(\partial x^1)^2} & \dfrac{\partial^2 f}{\partial x^1 \partial x^2} & \cdots & \dfrac{\partial^2 f}{\partial x^1 \partial x^n} \\ \dfrac{\partial^2 f}{\partial x^2 2x^1} & \dfrac{\partial^2 f}{(\partial x^2)^2} & \cdots & \dfrac{\partial^2 f}{\partial x^2 \partial x^n} \\ \vdots & \vdots & & \vdots \\ \dfrac{\partial^2 f}{\partial x^n \partial x^1} & \dfrac{\partial^2 f}{\partial x^n \partial x^2} & \cdots & \dfrac{\partial^2 f}{(\partial x^n)^2} \end{pmatrix}.$$

This matrix is called the _Hessian_ of the function $f : R^n \to R^1$.
Thus Newton's method for finding minima of functions is to
construct a sequence $\{x_k\}$ via the equations

$$\frac{\partial^2 f}{\partial x^2}(x_k)(x_{k+1} - x_k) = -\frac{\partial f}{\partial x}(x_k), \quad k = 0, 1, 2, \ldots .$$

Now the method is by no means guaranteed to work in
every case. First of all, as indicated in Theorem 5.3, the
initial point x_0 should lie fairly close to x^*. Also, Newton's

method pays no attention to the fact that we are looking for

a minimum of f - it is looking for a point where $\frac{\partial f}{\partial x} = 0$.

Such a point might or might not be a minimum. It might be a

maximum of f or a point where $\frac{\partial f}{\partial x} = 0$ but f has neither a

maximum nor a minimum. In fact, Newton's method for find-

ing maxima is exactly the same as Newton's method for find-

ing minima. Nevertheless, if these limitations are borne in

mind, the technique becomes extremely useful.

EXAMPLE. Two plastic balls, one with charge +1, the other

with charge -1, lie in parabolic troughs whose equations are

$$z = x^2, \qquad z = 2(x - 2)^2$$

respectively. (See Fig. 9). The two balls thus attract each

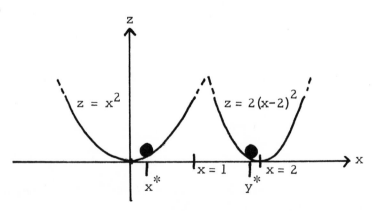

Figure 9

other but are hindered in moving toward each other by the

force of gravity. What are the equilibrium positions x^*, y^*

for the two balls?

A principle of mechanics states that the equilibrium

positions x^*, y^* are a minimum for the potential energy

function

$$V(x, y) = x^2 + 2(y - 2)^2 - \frac{1}{y - x} \, ,$$

where we have neglected any difference in height of the two

balls and have assumed convenient units. We have

$$\frac{\partial V}{\partial x} = 2x - \frac{1}{(y-x)^2} \, , \qquad \frac{\partial V}{\partial y} = 4(y-2) + \frac{1}{(y-x)^2}$$

$$\frac{\partial^2 V}{\partial x^2} = 2 - \frac{2}{(y-x)^3} \, , \qquad \frac{\partial^2 V}{\partial y^2} = 4 - \frac{2}{(y-x)^3}$$

$$\frac{\partial^2 V}{\partial x \partial y} = \frac{\partial^2 V}{\partial y \partial x} = \frac{2}{(y-x)^3} \, .$$

Thus Newton's method is

$$\begin{pmatrix} 2 - \dfrac{2}{(y_k - x_k)^3} & \dfrac{2}{(y_k - x_k)^3} \\[4mm] \dfrac{2}{(y_k - x_k)^3} & 4 - \dfrac{2}{(y_k - x_k)^3} \end{pmatrix} \begin{pmatrix} x_{k+1} - x_k \\[4mm] y_{k+1} - y_k \end{pmatrix} = - \begin{pmatrix} 2x_k - \dfrac{1}{(y_k - x_k)^2} \\[4mm] 4(y_k - 2) + \dfrac{1}{(y_k - x_k)^2} \end{pmatrix}$$

We begin with $x_0 = 0$, $y_0 = 2$. Thus

$$\begin{pmatrix} 2 - \frac{1}{4} & \frac{1}{4} \\ \frac{1}{4} & 4 - \frac{1}{4} \end{pmatrix} \begin{pmatrix} x_1 \\ y_1 - 2 \end{pmatrix} = - \begin{pmatrix} -\frac{1}{4} \\ \frac{1}{4} \end{pmatrix}$$

or

$$\frac{7}{4} x_1 + \frac{1}{4} (y_1 - 4) = \frac{1}{4}$$

$$\frac{1}{4} x_1 + \frac{15}{4} (y_1 - 4) = -\frac{1}{4} \ .$$

Solving we have $y_1 - 2 = -.077$, i. e. , $y_1 = 1.923$, and

$x_1 = .154.$

For the next step we have

$$\begin{pmatrix} 2 - \dfrac{2}{5.54} & \dfrac{2}{5.54} \\ \dfrac{2}{5.54} & 4 - \dfrac{2}{5.54} \end{pmatrix} \begin{pmatrix} x_2 - .154 \\ y_2 - 1.923 \end{pmatrix} - \begin{pmatrix} -.012 \\ .012 \end{pmatrix}$$

or

$$1.639 (x_2 - .154) + .361 (y_2 - 1.923) = .012$$

$$.361 (x_2 - .154) + 3.639 (y_2 - 1.923) = -.012 \ .$$

This gives $y_2 - 1.923 = -.004$, i. e. , $y_2 = 1.919$, and

$x_2 - .154 = .008$, i. e. , $x_2 = .162.$ Then

$$\begin{pmatrix} 2-\dfrac{2}{5.42} & \dfrac{2}{5.42} \\[2ex] \dfrac{2}{5.42} & 4-\dfrac{2}{5.42} \end{pmatrix} \begin{pmatrix} x_3 - .162 \\[2ex] y_3 - 1.919 \end{pmatrix} = \begin{pmatrix} .000 \\[2ex] .000 \end{pmatrix}$$

and we have gone as far as three decimal place accuracy can take us. Thus the equilibrium position is

$$x^* = .162$$

$$y^* = 1.919 .$$

We conclude with two remarks. In most standard numerical analysis texts it is shown that Newton's method for solving $g(x) = 0$ converges <u>quadratically</u>, i. e. ,

$$\|x_{k+1} - x^*\| \le K\|x_k - x^*\|^2 ,$$

for some $K > 0$ and sufficiently large k. This means that Newton's method, when it works, converges very rapidly to the desired solution x^*.

It is not absolutely necessary to compute $\dfrac{\partial g}{\partial x}(x_k)$ each time. Often the method

$$\frac{\partial g}{\partial x}(x_0)(x_{k+1} - x_k) = -g(x_k)$$

will also yield a sequence $\{x_k\}$ with $\lim_{k \to \infty} x_k = x^*$. Here, however, we have only <u>linear</u> convergence

$$\|x_{k+1} - x^*\| \le L\|x_k - x^*\|, \qquad 0 \le L < 1.$$

If this method is used one should compute the inverse matrix $(\frac{\partial g}{\partial x}(x_0))^{-1}$ and write

$$x_{k+1} = x_k - (\frac{\partial g}{\partial x}(x_0))^{-1} g(x_k).$$

EXERCISES.   CHAPTER 5

1. Compute the gradient, or Jacobian, as the case may be, for the following functions f:

   (i) $f : R^3 \to R^1$, $f(x, y, z) = \sqrt{x^2 + y^2 + z^2}$;

   (ii) $f : R^2 \to R^2$, $f^1(x, y) = \sin x + \cos y$,

   $$f^2(x,y) = \sin(x+y);$$

   (iii) $f : R^3 \to R^2$, $f^1(x, y, z) = \sqrt{x^2 + y^2 + z^2}$,

   $$f^2(x,y,z) = -x^3 + y^4 + z^5 + xyz;$$

   (iv) $f : R^1 \to R^3$, $f^1(x) = \sin x$,

   $$f^2(x) = \sqrt{1 - x^2},$$

   $$f^3(x) = \tan x.$$

2. Let A be a symmetric, positive definite $n \times n$ matrix. Show that for $x, y \in R^n$

$$(x'Ay)^2 \leq (x'Ax)(y'Ay).$$

When does equality hold?

(See proof of Schwartz Inequality, Lemma 5.1.)

3. Following the methods of Theorem 5.1, show that if $f : R^n \rightarrow R^1$ is continuously differentiable near x_0, then

$$f(x) = f(x_0) + \frac{\partial f}{\partial x}(x_0)(x - x_0)$$

$$+ \frac{1}{2}(x-x_0)'\frac{\partial^2 f}{\partial x^2}(x_0)(x-x_0) + (\|x-x_0\|^2), \quad x \rightarrow x_0.$$

4. Let $f : R^n \rightarrow R^1$ be given by

$$f(x) = x'Wx + b'x + c$$

where c is a constant, $b \in R^n$ and W is a (not necessarily symmetric) n ? n matrix. Compute the Hessian matrix $\dfrac{\partial^2 f}{\partial x^2}$.

5. Use Newton's method to find a point $\binom{x^*}{y^*} \in R^2$ which minimizes

$$\left\| \binom{x}{y} - \binom{0}{0} \right\|_e + \left\| \binom{x}{y} - \binom{1}{0} \right\|_e + \left\| \binom{x}{y} - \binom{1}{2} \right\|_e.$$

Does this example suggest to you a practical difficulty in using Newton's method?

6. A modification of Newton's method for solving $g(x) = 0$ is provided by

$$\frac{\partial g}{\partial x}(x_k)(x_{k+1} - x_k) = -\lambda g(x_k), \qquad k = 0, 1, 2, \dots .$$

(i) For which values of $\lambda > 0$ will this method work to solve the linear equation

$$g(x) = Ax + b = 0 ,$$

where A is a non-singular $n \times n$ matrix, i. e. for which values of $\lambda > 0$ do we have

$$\lim_{k \to \infty} x_k = -A^{-1}b ?$$

(ii) Look at the example following Theorem 5.3 and see if the modified Newton's method will solve the equation given there for some value of λ .

7. Let A be an $n \times n$ matrix. Show that

(i) $\|A\|_s = \max_{j=1,2,\dots,n} \{ \sum_{i=1}^{n} |a_i^j| \};$

(ii) $\|A\|_t = \max_{i=1,2,\dots,n} \{ \sum_{j=1}^{n} |a_i^j| \}.$

CHAPTER 6 CONVEXITY

In Chapter 5 we proved that if a function $f : R^n \to R^1$ assumes a minimum in a set S at a point x^* in the interior of S, then $\frac{\partial f}{\partial x}(x^*) = 0$. As we pointed out, however, this is only a necessary condition. The mere fact that $\frac{\partial f}{\partial x}$ vanishes at some point is not sufficient to guarantee that f has either a maximum or a minimum at that point. In a search for sufficient condition to guarantee maxima or minima we are led to a class of functions called <u>convex functions</u> and a class of sets called <u>convex sets.</u>

Recall that if x_1 and x_2 are points in R^n then the point

$$x(\lambda) = \lambda x_2 + (1 - \lambda)x_1, \qquad 0 \le \lambda \le 1$$

183

lies on the straight line segment joining x_1 and x_2 with

$$x(0) = x_1, \ x(1) = x_2.$$

DEFINITION 6.1.   A set $S \subseteq R^n$ is <u>convex</u> if whenever

$x_1, x_2 \in S$, then

$$x(\lambda) = \lambda x_2 + (1 - \lambda)x_1 \in S, \qquad 0 \le \lambda \le 1.$$

The set S is <u>strictly convex</u> if whenever $x_1 \neq x_2$, $x_1, x_2 \in \overline{S}$,

the closure of S, then $x(\lambda) \in S^0$, the interior of S, for

$0 < \lambda < 1$.   Clearly a strictly convex set is also convex.

EXAMPLES.   Consider the sets

$$S_1 = \{ \begin{pmatrix} x \\ y \end{pmatrix} \in R^2 \, | \, x^2 + y^2 \le 1 \},$$

$$S_2 = \{ \begin{pmatrix} x \\ y \end{pmatrix} \in R^2 \, | \, |x| + |y| \le 1 \},$$

$$S_3 = \{ \begin{pmatrix} x \\ y \end{pmatrix} \in R^2 \, | \, |x|^{\frac{1}{2}} + |y|^{\frac{1}{2}} \le 1 \},$$

as shown in Figure 10. S_1 is strictly convex, for the interior

of the line segment joining any two points of its closure lies

in its interior. S_2 is convex but not strictly convex.   The

line segment joining the point $\begin{pmatrix} 0 \\ 1 \end{pmatrix}$ to the point $\begin{pmatrix} 1 \\ 0 \end{pmatrix}$ lies

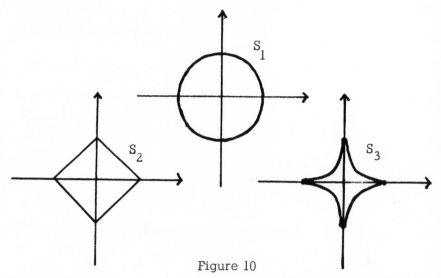

Figure 10

entirely in ∂S_2. S_3 is not convex for the line segment

joining $\binom{0}{1}$ to $\binom{1}{0}$ lies wholly outside S_3 except for the

points $\binom{0}{1}$ and $\binom{1}{0}$.

DEFINITION 6.2. Let $f : R^n \to R^1$ be defined on a domain D

which is a convex set. We say that f is a <u>convex function</u>

if whenever $x_1, x_2 \in D$ then, for $0 \le \lambda \le 1$,

$$f(x(\lambda)) = f(\lambda x_2 + (1-\lambda)x_1) \le \lambda f(x_2) + (1-\lambda)f(x_1).$$

If this inequality holds strictly, i.e.

$$f(x(\lambda)) < \lambda f(x_2) + (1-\lambda)f(x_1)$$

when $x_1 \neq x_2$, $0 < \lambda < 1$, then we say that f is a <u>strictly</u>

convex function. If $-f$ is a convex function (strictly convex

function) we say that f is a <u>concave</u> function (<u>strictly con-

cave function</u>).

The graph of a convex function is, roughly speaking,

bowl shaped, the point $(x(\lambda), f(x(\lambda)))$ always lies on or be-

low the straight line segment joining the points $(x_1, f(x_1))$,

$(x_2, f(x_2))$ in R^{n+1}. Some simple examples are shown in

Figure 11.

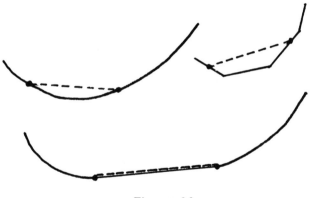

Figure 11

The following theorem also provides some examples.

THEOREM 6.1. Let $f : R^n \to R^1$ be given by

$$f(x) = x'Ax + b'x + c,$$

where A is a symmetric $n \times n$ matrix, $b \in R^n$ and c is a real constant. Then

(i)   if A is positive, f is convex;

(ii)   if A is positive definite, f is strictly convex;

(iii) if A is negative (- A is positive), f is concave;

(iv)   if A is negative definite (- A is positive definite),

f is strictly concave.

<u>Proof.</u> Clearly it is enough to prove (i) and (ii). Let A be

positive and compute

$$[\lambda f(x_2) + (1-\lambda)f(x_1)] - [f(\lambda x_2 + (1-\lambda)x_1)]$$

$$= \lambda[x_2'Ax_2 + b'x_2 + c] + (1-\lambda)[x_1'Ax_1 + b'x_1 + c]$$

$$- [\lambda x_2 + (1-\lambda)x_1]'A[\lambda x_2 + (1-\lambda)x_1] - b'[\lambda x_2 + (1-\lambda)x_1] - c$$

$$= \lambda x_2'Ax_2 + (1-\lambda)x_1'Ax_1 - [\lambda x_2 + (1-\lambda)x_1]'A[\lambda x_2 + (1-\lambda)x_1]$$

$$= (\lambda-\lambda^2)x_2'Ax_2 - 2\lambda(1-\lambda)x_2'Ax_1 + ((1-\lambda)-(1-\lambda)^2)x_1'Ax_1$$

$$= \lambda(1-\lambda)[x_2'Ax_2 - 2x_2'Ax_1 + x_1'Ax_1]$$

$$= \lambda(1-\lambda)(x_2-x_1)'A(x_2-x_1) \geq 0, \qquad 0 \leq \lambda \leq 1.$$

Hence f is convex. If A is positive definite, $x_1 \neq x_2$ and

$0 < \lambda < 1$ then clearly

$$\lambda (1 - \lambda)(x_2 - x_1)' A(x_2 - x_1) > 0$$

and thus f is strictly convex if A is positive definite. This completes the proof.

THEOREM 6. 2.   If $f : R^n \to R^1$ is convex in the open set S then f is continuous in S.

Proof. We will give the proof only for n = 1. The proof in the general case makes use of essentially the same ideas but the additional complexity is too great to warrant its presentation here.

Let r_0 be a real number in S and let r_1 and r_2 be such that $r_1 < r_0 < r_2$. Define linear functions $g_1(r) =$
$$\left(\frac{r_0 - r}{r_0 - r_1}\right) f(r_1) + \left(\frac{r - r_1}{r_0 - r_1}\right) f(r_0), \quad g_2(r) = \left(\frac{r - r_0}{r_2 - r_0}\right) f(r_2) +$$
$$\left(\frac{r_2 - r}{r_2 - r_0}\right) f(r_0), \quad -\infty < r < \infty.$$

Let r lie between r_1 and r_0. Then the convexity of f implies that

$$f(r) \leq \left(\frac{r_0 - r}{r_0 - r_1}\right) f(r_1) + \left(\frac{r - r_1}{r_0 - r_1}\right) f(r_0) = g_1(r)$$

since $\quad r = \left(\frac{r_0 - r}{r_0 - r_1}\right) r_1 + \left(\frac{r - r_1}{r_0 - r_1}\right) r_0 .$

On the other hand r_0 lies between r and r_2,

$$r_0 = \frac{r_2 - r_0}{r_2 - r} r + \frac{r_0 - r}{r_2 - r} r_2 ,$$

and therefore, again by convexity of f,

$$f(r_0) \leq \frac{r_2 - r_0}{r_2 - r} f(r) + \frac{r_0 - r}{r_2 - r} f(r_2)$$

so that

$$f(r) \geq \frac{r_2 - r}{r_2 - r_0} \left[f(r_0) - \frac{r_0 - r}{r_2 - r} f(r_2) \right]$$

$$= \frac{r_2 - r}{r_2 - r_0} f(r_0) + \frac{r - r_0}{r_2 - r_0} f(r_2) = g_2(r) .$$

Thus

$$g_2(r) \leq f(r) \leq g_1(r) , \qquad r_1 \leq r \leq r_0 .$$

Similar arguments show that

$$g_1(r) \leq f(r) \leq g_2(r) , \qquad r_0 \leq r \leq r_2 .$$

Then since $\lim_{r \to r_0} g_1(r) = \lim_{r \to r_0} g_2(r)$ we conclude that

$\lim_{r \to r_0} f(r) = f(r_0)$ and thus f is continuous at r_0 and the

proof is complete.

The next theorem establishes an important relation-

ship between convex functions and convex sets.

THEOREM 6. 3.   If $f : R^n \to R^1$ is a convex function defined

on a convex domain D then the set

$$S_d = \{x \in D \mid f(x) \le d\}$$

is a convex subset of D.   If D and f are strictly convex,

so is S_d.

Proof.   Let $x_1, x_2 \in S_d$ and put $x(\lambda) = \lambda x_2 + (1-\lambda)x_1$.   Then,

for $0 \le \lambda \le 1$,

$$f(x(\lambda)) \le \lambda f(x_2) + (1-\lambda)f(x_1) \le \lambda d + (1-\lambda)d = d$$

and it follows that $x(\lambda) \in S_d$.   Thus S_d includes the line

segment between any two of its points and is convex.

Now let x_1 and x_2 be distinct points in S_d and let

$0 < \lambda < 1$. Since D is strictly convex the point $x(\lambda)$ lies in D^0. Therefore, for some $\delta_1 > 0$, $x \in D$ if $\|x-x(\lambda)\| < \delta_1$. Also

$$f(x(\lambda)) < \lambda f(x_2) + (1-\lambda)f(x_1) \leq d$$

and, since f is continuous, there is a $\delta_2 > 0$ such that $f(x) \leq d$ if $\|x-x(\lambda)\| < \delta_2$. Letting $\delta = \min\{\delta_1, \delta_2\}$ we see that $x \in S_c$ is $\|x-x(\lambda)\| < \delta$. Thus, if $x_1 \neq x_2$, $0 < \lambda < 1$, the point $x(\lambda)$ lies in S_d^0 and S_d is strictly convex. This completes the proof.

Important examples are provided by the sets

$$S_d = \{x \in R^n \mid x'Ax + b'x + c \leq d\}$$

which are convex or strictly convex according as the matrix A is positive or positive definite, respectively.

We now present a result which provides conditions under which a function f has a unique minimum in a set S.

THEOREM 6.4. Let f be a convex function defined on a domain D which includes a convex set S. If there exists a point $x^* \in S$ such that $f(x) \geq f(x^*)$ for all $x \in S$ then the set

$$M = \{y \in S \mid f(y) = f(x^*)\}$$

is a convex subset of S. If there is a spherical neighborhood

of x^* in which f is strictly convex, then x^* is the only

point in M and thus f achieves its minimum in S only at

the point x^*.

Proof. Theorem 6.3, with D replaced by S, shows that the

set

$$\tilde{M} = \{y \in S \mid f(y) \leq f(x^*)\}$$

is convex. But $f(y) < f(x^*)$ is impossible for $y \in S$ and we

conclude $\tilde{M} = M$ and thus M is convex.

Now assume f is strictly convex in $N(x^*, \delta) =$

$\{x \in R^n \mid \|x - x^*\| < \delta\}$ for some $\delta > 0$ and let y be a point

in S different from x^*. (If there is no such point there is

nothing to prove.) Put $x(\lambda) = \lambda y + (1-\lambda)x^*$ and note that

$x(\lambda) \in N(x^*, \delta)$ for $0 \leq \lambda < \lambda_1$ for some λ_1 satisfying

$0 < \lambda_1 < 1$. Since f is convex, we have

$$f(x(\lambda)) \leq \lambda f(y) + (1 - \lambda)f(x^*)$$

and thus

$$f(y) \geq \frac{1}{\lambda} [f(x(\lambda)) - (1 - \lambda)f(x^*)] .$$

Now $x(\lambda)$ and x^* both lie in the convex set $N(x^*, \delta)$ if $0 \leq \lambda < \lambda_1$. Assuming λ is thus restricted, we put

$$\hat{x} = \tfrac{1}{2} x(\lambda) + \tfrac{1}{2} x^* .$$

Since f is strictly convex in $N(x^*, \delta)$ which includes both $x(\lambda)$ and x^* we have

$$f(\hat{x}) < \tfrac{1}{2} f(x(\lambda)) + \tfrac{1}{2} f(x^*)$$

whence

$$f(x(\lambda)) > 2[f(\hat{x}) - \tfrac{1}{2} f(x^*)] .$$

But $f(\hat{x}) \geq f(x^*)$ so

$$f(x(\lambda)) > 2[f(x^*) - \tfrac{1}{2} f(x^*)] = f(x^*) .$$

Then, returning to our earlier inequality for $f(y)$,

$$f(y) \geq \frac{1}{\lambda} [f(x(\lambda)) - (1 - \lambda)f(x^*)]$$
$$> \frac{1}{\lambda} [f(x^*) - (1 - \lambda)f(x^*)] = f(x^*)$$

and thus $f(y) > f(x^*)$ for all $y \in S$ with $y \neq x^*$. With this the proof is complete.

COROLLARY. Let f be a concave function defined on a domain D which includes a convex set S. If there exists a point $x^* \in S$ such that $f(x) \leq f(x^*)$ for all $x \in S$ then the set

$$M = \{y \in S \mid f(y) = f(x^*)\}$$

is a convex subset of S. If there is a spherical neighborhood of x^* in which f is strictly concave then x^* is the only point in M and thus f achieves its maximum in S only at the point x^*.

Proof. Just replace f by $-f$ and use Theorem 6.4.

We now turn to the proof of some theorems which relate the convexity of a function $f : R^n \to R^1$ to the gradient of f, $\frac{\partial f}{\partial x}$, and the Hessian of f, $\frac{\partial^2 f}{\partial x^2}$, when these are defined. A by-product of the first theorem will be that the condition $\frac{\partial f}{\partial x}(x^*) = 0$ is sufficient for f to assume a minimum at x^* if f is a convex function.

THEOREM 6. 5.   Let $f : R^n \to R^1$ be a convex function defined

on a convex domain D and continuously differentiable near

$x_0 \in D$.   Let y be any point in D.   Then

$$f(y) \geq f(x_0) + \frac{\partial f}{\partial x}(x_0)(y - x_0)$$

and if f is strictly convex and $y \neq x_0$ the inequality holds

strictly.

REMARK.   As indicated in Figure 12, this result shows that a

convex function is everywhere greater than or equal to its

linear approximation at x_0.

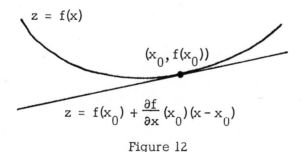

Figure 12

<u>Proof of Theorem 6. 5.</u>   We will obtain a proof by contradiction.

Let us suppose

$$f(y) = f(x_0) + \frac{\partial f}{\partial x}(x_0)(y - x_0) - \delta, \quad \delta > 0.$$

Let $x(\lambda) = \lambda y + (1 - \lambda)x_0$ and compute

$$f(x(\lambda)) = f(x_0) + \frac{\partial f}{\partial x}(x_0)(x(\lambda) - x_0) + \theta(\|x(\lambda) - x_0\|)$$

$$= f(x_0) + \frac{\partial f}{\partial x}(x_0)(\lambda(y - x_0)) + \theta(\lambda\|y - x_0\|)$$

$$= f(x_0) + \lambda\frac{\partial f}{\partial x}(x_0)(y - x_0) + \theta(\lambda)\|y - x_0\|, \quad \lambda \to 0.$$

Then, using $f(y) = f(x_0) + \frac{\partial f}{\partial x}(x_0)(y - x_0) - \delta$, we have

$$f(x(\lambda)) - [\lambda f(y) + (1 - \lambda)f(x_0)]$$

$$= f(x_0) + \lambda\frac{\partial f}{\partial x}(x_0)(y - x_0) + \theta(\lambda)\|y - x_0\|$$

$$- [\lambda(f(x_0) + \frac{\partial f}{\partial x}(x_0)(y - x_0) - \delta) + (1 - \lambda)f(x_0)]$$

$$= \lambda\delta + \theta(\lambda)\|y - x_0\|$$

which is positive for positive λ sufficiently close to zero.
Thus for small $\lambda > 0$

$$f(x(\lambda)) > [\lambda f(y) + (1 - \lambda)f(x_0)],$$

contradicting the convexity of f. Thus we must have

$$f(y) \geq f(x_0) + \frac{\partial f}{\partial x}(x_0)(y - x_0).$$

Now suppose f is strictly convex. We wish to prove

$$f(y) > f(x_0) + \frac{\partial f}{\partial x}(x_0)(y - x_0)$$

if $y \neq x_0$. We will suppose that this is false, i.e., that

$$f(y) = f(x_0) + \frac{\partial f}{\partial x}(x_0)(y - x_0).$$

$(f(y) < f(x_0) + \frac{\partial f}{\partial x}(x_0)(y - x_0)$ is ruled out by our first result.) Since f is strictly convex and $y \neq x_0$, if we put $x_1 = \frac{y + x_0}{2}$ then

$$f(x_1) < \tfrac{1}{2} f(y) + \tfrac{1}{2} f(x_0) = f(x_0) + \tfrac{1}{2} \frac{\partial f}{\partial x}(x_0)(y - x_0)$$

$$= f(x_0) + \frac{\partial f}{\partial x}(x_0)(\frac{y - x_0}{2}) = f(x_0) + \frac{\partial f}{\partial x}(x_0)(x_1 - x_0)$$

and this contradicts the convexity of f again. So we must have $f(y) > f(x_0) + \frac{\partial f}{\partial x}(x_0)(y - x_0)$. This completes the proof.

Now we present our result on the sufficiency of the condition $\frac{\partial f}{\partial x}(x^*) = 0$.

THEOREM 6.6. If $f : R^n \to R^1$ is a convex function defined in a convex domain D and continuously differentiable near

$x_0 \in D$ then the condition $\frac{\partial f}{\partial x}(x^*) = 0$ implies $f(y) \geq f(x^*)$

for all $y \in D$. If f is strictly convex and $y \neq x^*$ we have

$f(y) > f(x^*)$.

Proof. Just replace x_0 by x^* in Theorem 6.5 and we have

$$f(y) \geq f(x^*) + 0(y - x^*) = f(x^*)$$

and this inequality holds strictly if f is strictly convex and

$y \neq x^*$.

The next result is a converse to Theorem 6.5.

THEOREM 6.7. If $f : R^n \to R^1$ is defined on a convex domain

D and continuously differentiable there the inequality

$$f(y) \geq f(x_0) + \frac{\partial f}{\partial x}(x_0)(y - x_0),$$

valid for all $y, x_0 \in D$, implies f is convex in D. If this

inequality holds strictly whenever $y \neq x_0$ then f is strictly

convex in D.

Proof. Let x_1 and x_2 be any points in D and let

$x(\lambda) = \lambda x_2 + (1-\lambda)x_1$, $0 \leq \lambda \leq 1$. From our hypothesis we have

$$f(x_1) \geq f(x(\lambda)) + \frac{\partial f}{\partial x}(x(\lambda))(x_1 - x(\lambda))$$

$$f(x_2) \geq f(x(\lambda)) + \frac{\partial f}{\partial x}(x(\lambda))(x_2 - x(\lambda)).$$

But $x_2 - x(\lambda) = \frac{\lambda-1}{\lambda}(x_1 - x(\lambda))$ so

$$f(x_2) \geq f(x(\lambda)) + \frac{\lambda-1}{\lambda}\frac{\partial f}{\partial x}(x(\lambda))(x_1 - x(\lambda)).$$

Multiplying the first of these three inequalities by $(1-\lambda)$ and the third by λ and adding we obtain

$$(1-\lambda)f(x_1) + \lambda f(x_2) \geq f(x(\lambda))$$

and thus f is convex in D. If $x_1 \neq x_2$ and $0 < \lambda < 1$ each of the above inequalities is immediately seen to hold strictly and f is strictly convex in D. This completes the proof.

We now refer the reader to the definition of the Hessian matrix given in Chapter 5 and to Exercise 3 of Chapter 5 which states that if f is twice continuously differentiable near x_0 then

$$f(y) = f(x_0) + \frac{\partial f}{\partial x}(x_0)(y-x_0) + \tfrac{1}{2}(y-x_0)' \frac{\partial^2 f}{\partial x^2}(x_0)(y-x_0)$$

$$+ \mathscr{O}(\|y \ x_0\|^2),$$

valid as $y \to x_0$. The matrix $\dfrac{\partial^2 f}{\partial x^2}$ has entries

$$\left(\frac{\partial^2 f}{\partial x^2}\right)^i_j = \frac{\partial^2 f}{\partial x^j \partial x^i}$$

and is consequently symmetric, for it is well known that if f is twice continuously differentiable then $\dfrac{\partial^2 f}{\partial x^j \partial x^i} = \dfrac{\partial^2 f}{\partial x^i \partial x^j}$.

We will now see that convexity is related to the positivity of this symmetric matrix.

THEOREM 6.8. Let $f : R^n \to R^1$ be twice continuously differentiable in an open domain D. Then f is convex if and only if the Hessian $\dfrac{\partial^2 f}{\partial x^2}$ is everywhere positive in D. If $\dfrac{\partial^2 f}{\partial x^2}$ is everywhere positive definite in D then f is strictly convex in D.

Proof. Suppose there were a point $x_0 \epsilon D$ and a non-zero vector $y \epsilon R^n$ such that

$$y' \frac{\partial^2 f}{\partial x^2}(x_0)y = -\delta < 0 .$$

Since D is open the points $x_1 = x_0 - \mu y$, $x_2 = x_0 + \mu y$ belong to D for small $\mu > 0$. We compute

$$f(x_j) = f(x_0) + \frac{\partial f}{\partial x}(x_0)(x_j - x_0) + \tfrac{1}{2}(x_j - x_0)' \frac{\partial^2 f}{\partial x^2}(x_0)(x_j - x_0)$$

$$+ \mathcal{O}(\|x_j - x_0\|^2),$$

for $j = 1, 2$. Since $x_j - x_0 = (-1)^j \mu y$

$$f(x_j) = f(x_0) + (-1)^j \mu \frac{\partial f}{\partial x}(x_0)y + \tfrac{1}{2}\mu^2 y' \frac{\partial^2 f}{\partial x^2}(x_0)y + \mathcal{O}(\mu^2) .$$

Adding the two equations for $j = 1, 2$ we have

$$f(x_1) + f(x_2) = 2f(x_0) - \tfrac{1}{2}\mu^2 \delta + \mathcal{O}(\mu^2)$$

from which we conclude that

$$\tfrac{1}{2}f(x_1) + \tfrac{1}{2}f(x_2) < f(x_0)$$

if μ is sufficiently small. But since $x_0 = \tfrac{1}{2}x_1 + \tfrac{1}{2}x_2$ this
contradicts the convexity of f. Therefore, if f is convex,
$\frac{\partial^2 f}{\partial x^2}$ is positive, i. e. $y' \frac{\partial^2 f}{\partial x^2}(x_0)y \geq 0$ for all $x_0 \in D$, $y \neq 0$.

Now suppose $\dfrac{\partial^2 f}{\partial x^2}$ is everywhere positive in D. Let

x_0 and y lie in D and let $x(\lambda) = \lambda y + (1-\lambda)x_0$. Put

$$g(\lambda) = f(x(\lambda))$$

so that $g(0) = f(x_0)$, $g(1) = f(y)$. We have already seen in

Chapter 5 that

$$\frac{dg}{d\lambda} = \frac{\partial f}{\partial x}(x(\lambda))(y - x_0).$$

Let us put $h(\lambda) = \dfrac{\partial f}{\partial x}(x(\lambda))(y - x_0) = (y-x_0)'\dfrac{\partial f}{\partial x}(x(\lambda))'$ and

compute

$$\frac{d^2 g}{d\lambda^2} = \frac{dh}{d\lambda} = \frac{d}{d\lambda}\left((y-x_0)'\frac{\partial f}{\partial x}(x(\lambda))'\right)$$

$$= \frac{d}{d\lambda}\left(\sum_{i=1}^{n}(y^i - x_0^i)\frac{\partial f}{\partial x^i}(x(\lambda))\right)$$

$$= \sum_{i=1}^{n}(y^i - x_0^i)\frac{\partial}{\partial x}\left(\frac{\partial f}{\partial x^i}\right)(x(\lambda))\frac{dx(\lambda)}{d\lambda}$$

$$= \sum_{i=1}^{n}(y^i - x_0^i)\frac{\partial}{\partial x}\left(\frac{\partial f}{\partial x^i}\right)(x(\lambda))(y - x_0)$$

$$= (y - x_0)'\frac{\partial^2 f}{\partial x^2}(x(\lambda))(y - x_0).$$

Now the fact that $\dfrac{\partial^2 f}{\partial x^2}(x(\lambda))$ is positive shows that

$$\frac{d^2 g}{d\lambda^2} = (y - x_0)' \frac{\partial^2 f}{\partial x^2}(x(\lambda))(y - x_0) \geq 0.$$

Then, for some $\hat{\lambda}$ with $0 < \hat{\lambda} < 1$,

$$g(1) = g(0) + \frac{dg}{d\lambda}(\hat{\lambda})$$

$$= g(0) + \frac{dg}{d\lambda}(0) + \int_0^{\hat{\lambda}} \frac{d^2 g}{d\lambda^2} d\lambda \geq g(0) + \frac{dg}{d\lambda}(0)$$

whence

$$f(y) \geq f(x_0) + \frac{\partial f}{\partial x}(x_0)(y - x_0)$$

and f is convex. If $\dfrac{\partial^2 f}{\partial x^2}$ is positive definite everywhere in

D then $\dfrac{d^2 g}{d\lambda^2} = (y - x_0)' \dfrac{\partial^2 f}{\partial x^2}(x(\lambda))(y - x_0) > 0$ when $y \neq x_0$ and

we obtain

$$f(y) > f(x_0) + \frac{\partial f}{\partial x}(x_0)(y - x_0), \qquad y \neq x_0,$$

and f is strictly convex, thus completing the proof.

The condition that D be open in Theorem 6.8 cannot

be removed, at least as far as the result that convexity implies positivity of the Hessian is concerned.   Consider the example in R^2

$$f(x, y) = xy$$

$$D = \{(\begin{smallmatrix} x \\ y \end{smallmatrix}) \mid x = y\}.$$

Then $f(x, y) = x^2$ is convex in D but the Hessian matrix is everywhere $(\begin{smallmatrix} 0 & 1 \\ 1 & 0 \end{smallmatrix})$, which is not positive.

A function f may be strictly convex without the Hessian being everywhere positive definite.   The function $f(x, y) = x^4 + y^4$ is strictly convex but the Hessian is $(\begin{smallmatrix} 0 & 0 \\ 0 & 0 \end{smallmatrix})$ at the origin and thus not positive definite there.

EXERCISES   CHAPTER 6

1.   Let $f : R^1 \to R^1$ have the properties

$$f(0) = 0$$

$$f(r) \geq 0, \ r \geq 0$$

$$f \text{ is convex for } r \geq 0.$$

Put $g(x, y) = f(x^2 + y^2)$.   Show that $g : R^2 \to R^1$ is convex.

2. Let $f : R^n \to R^1$ be non-negative for all x. Show that the square of f, f^2, is also convex. Show that this is not necessarily true if we allow f to assume negative values.

3. Let $x_1, x_2, \ldots, x_r$ be points in R^n. Consider two problems:

P_1: Find $x^* \in R^n$ such that x^* is a minimum for

$$f(x) = \sum_{i=1}^{r} \|x - x_i\|_e .$$

P_2: Same as P_1 but

$$f(x) = \sum_{i=1}^{r} \|x - x_i\|_e^\alpha, \quad \alpha > 1.$$

Discuss the uniqueness of x^* in each case.

4. Let $f : R^n \to R^1$ and let S be the subset of R^{n+1} given by

$$S = \{ \binom{x}{y}, \ y \ \text{real}, \ x \in R^n \ | \ y \geq f(x) \}.$$

Show that f is convex if and only if S is convex and f is strictly convex if and only if S is strictly convex.

5. Let S be a strictly convex set in R^n and let $f : R^n \to R^1$

be differentiable and convex in S with the property that
$\frac{\partial f}{\partial x} \neq 0$ in S. Show that there is at most one point $x^* \in S$
such that $f(x^*) \leq f(x)$ for all $x \in S$.

6.   Let the set S be described by

$$S = \{x \in R^n \mid f(x) \leq c\}$$

where $f : R^n \to R^1$ is convex and differentiable and
$\frac{\partial f}{\partial x}(x_0) \neq 0$ whenever $f(x_0) = c$. Let x_0 be a point such
that $f(x_0) = c$ and let $v = \frac{\partial f}{\partial x}(x_0)'$. Let λ be any
positive number and put $x_1 = x_0 + \lambda v$. Show that

$$\|x_1 - x_0\|_e < \|x_1 - x\|_e$$

whenever x is a point of S different from x_0.

7.   Two islands have shapes describable by

$$f(x,y) \leq c_1, \quad g(x,y) \leq c_2$$

where f and g are continuously differentiable strictly
convex functions whose gradient vectors are non-zero
when $f(x,y) = c_1$, $g(x,y) = c_2$, respectively. It is
desired to build a bridge between the two islands of

shortest possible length.   Using the result of Exercise 6,

devise a search technique to find the optimal bridge lo-

cation.   Test your routine on

$$f(x,y) = x^2 + 2y^2, \quad c_1 = 1$$

$$g(x,y) = (x-4)^4 + (y-3)^2, \quad c_2 = 1.$$

CHAPTER 7.   BASES AND EIGENVECTORS

We begin this chapter by introducing the fundamental notions of linear dependence and linear independence of vectors.

DEFINITION 7.1.   Let $x_1, x_2, \ldots, x_r$ be vectors in the space R^n. We say that these vectors are <u>linearly dependent</u> if there are real numbers $\alpha^1, \alpha^2, \ldots, \alpha^r$, not all zero, such that

$$\alpha^1 x_1 + \alpha^2 x_2 + \ldots + \alpha^r x_r = 0.$$

We say that these vectors are <u>linearly independent</u> if the above equation implies $\alpha^1 = \alpha^2 = \ldots = \alpha^r = 0$.

Letting $x_i = \begin{pmatrix} x_i^1 \\ \vdots \\ x_i^n \end{pmatrix}$, $i = 1, 2, \ldots, r$, the above vector equation becomes a system of n equations in the r un-

knowns $\alpha^1, \alpha^2, \ldots, \alpha^r$:

$$\alpha^1 x_1^1 + \alpha^2 x_2^1 + \ldots + \alpha^r x_r^1 = 0$$

$$\alpha^1 x_1^2 + \alpha^2 x_2^2 + \ldots + \alpha^r x_r^2 = 0$$

$$\vdots \qquad \vdots \qquad \qquad \vdots$$

$$\alpha^1 x_1^n + \alpha^2 x_2^n + \ldots + \alpha^r x_r^n = 0 \ .$$

Letting X be the $n \times r$ matrix with entries x_i^j, $j = 1,2,\ldots,n$,

$i = 1, 2, \ldots, r$, and letting α be the column vector with com-

ponents α^i, $i = 1, 2, \ldots, r$, this system of equations be-

comes the vector equation

$$X\alpha = 0 .$$

If $r = n$ we know that this equation has a solution

$\alpha \neq 0$ if and only if $\det X = 0$.

If $r > n$, we augment the matrix X by adding $r - n$

zero rows, thereby obtaining an $r \times r$ matrix $\hat{X}$ with com-

ponents $\hat{x}_i^j = x_i^j$, $1 \leq j \leq n$, $\hat{x}_i^j = 0$, $n+1 \leq j \leq r$. Then the

equation $X\alpha = 0$ is clearly equivalent to

$$\hat{X}\alpha = 0 .$$

But $\hat{X}$, having at least one zero row, has determinant equal to zero. So in this case we conclude that there is a solution $\alpha \neq 0$.

If $r < n$ we proceed as follows. Let $\hat{X}$ be an $r \times r$ matrix formed from r rows of X. Suppose for some choice of r rows of X we have $\det \hat{X} \neq 0$. The equation $X\alpha = 0$ implies $\hat{X}\alpha = 0$. But $\hat{X}\alpha = 0$ only if $\alpha = 0$. So in this case $X\alpha = 0$ has only the solution $\alpha = 0$. It can be shown (we will not do it here) that if no such $\hat{X}$ with $\det \hat{X} \neq 0$ exists then $X\alpha = 0$ has a non-zero solution. Thus we have proved

THEOREM 7.1.   Let $x_1, x_2, \ldots, x_r$ be vectors in R^n. Then

(i)   If $r = n$, $x_1, x_2, \ldots, x_n$ are linearly dependent if
        $\det X = 0$ and linearly independent if $\det X \neq 0$;

(ii)   If $r > n$, $x_1, x_2, \ldots, x_r$ are linearly dependent;

(iii)   If $r < n$, $x_1, x_2, \ldots, x_r$ are linearly dependent if there
        do not exist r rows of X forming an $r \times r$ matrix $\hat{X}$
        with $\det \hat{X} \neq 0$. If such a matrix $\hat{X}$ does exist,
        $x_1, x_2, \ldots, x_r$ are linearly independent.

DEFINITION 7.2. A set of n linearly independent vectors in R^n is called a <u>basis for</u> R^n.

THEOREM 7.2. Let $x_1, x_2, \ldots, x_n$ be a basis for R^n and let $x \in R^n$. Then there are uniquely determined real numbers $\alpha^1, \alpha^2, \ldots, \alpha^n$ such that

$$x = \alpha^1 x_1 + \alpha^2 x_2 + \ldots + \alpha^n x_n .$$

REMARK. $\alpha^1, \alpha^2, \ldots, \alpha^n$ are called the <u>components</u> of x relative to the basis $x_1, x_2, \ldots, x_n$.

<u>Proof.</u> The vectors $x_1, x_2, \ldots, x_n$ form a basis for R^n if and only if $\det X \neq 0$. But in this case the inhomogeneous equation

$$X \alpha = x$$

has exactly one solution, which completes the proof.

Any subset U of R^n which has the property that $\alpha x + \beta y \in U$ whenever α and β are real numbers and x and y belong to U is called a subspace of R^n. For example,

$$U = \left\{ x = \begin{pmatrix} x^1 \\ x^2 \\ x^3 \end{pmatrix} \in R^3 \mid x^3 = 0 \right\}$$

is a subspace of R^3. In fact any plane in R^3 containing the origin or any line in R^3 containing the origin will be a subspace of R^3.

DEFINITION 7.3. Let U be a subspace of R^n. We define the _dimension_ of U, dim U, to be the largest integer r such that U contains a set $x_1, x_2, \ldots, x_r$ of linearly independent vectors.

From Theorem 7.1 we see that taking $U = R^n$ we get dim $R^n = n$, as we would expect. It is clear that if U is a subspace of R^n then dim $U \leq n$.

If $x_1, x_2, \ldots, x_r$ are vectors in R^n, then

$$U = \{ x \in R^n \mid x = \alpha^1 x_1 + \alpha^2 x_2 + \ldots + \alpha^r x_r, \alpha^1, \alpha^2, \ldots, \alpha^r \text{ real} \}$$

is clearly a subspace of R^n. The vectors $x_1, x_2, \ldots, x_r$ are called _generators_ of R^n. We see then that a basis for R^n is a set of n linearly independent generators of R^n. For subspaces U of R^n we have the following result.

THEOREM 7.3. Let U be an r-dimensional subspace of R^n.

Then any set of r linearly independent vectors in U generates U and hence constitutes a <u>basis</u> for U.

<u>Proof</u>. Let $x_1, x_2, \ldots, x_r$ be r linearly independent vectors in R^n. Let x be any vector in R^n. Then, by definition of $\dim U$, the set of vectors $x, x_1, x_2, \ldots, x_r$ must be linearly dependent. Consequently, there are real numbers $\alpha^0, \alpha^1, \alpha^2, \ldots, \alpha^r$, not all zero, such that

$$\alpha^0 x + \alpha^1 x_1 + \alpha^2 x_2 + \ldots + \alpha^r x_r = 0.$$

Now $\alpha^0 \neq 0$, otherwise $\alpha^1, \alpha^2, \ldots, \alpha^r$ are not all zero and $\alpha^1 x_1 + \alpha^2 x_2 + \ldots + \alpha^r x_r = 0$, contradicting the linear independence of $x_1, x_2, \ldots, x_r$. Then, dividing by α_0,

$$x = -\frac{\alpha_1}{\alpha_0} x_1 - \frac{\alpha_2}{\alpha_0} x_2 - \ldots - \frac{\alpha_r}{\alpha_0} x_r.$$

Thus every $x \in U$ has the form $x = \beta^1 x_1 + \beta^2 x_2 + \ldots + \beta^r x_r$ so that $x_1, x_2, \ldots, x_r$ generate U. Moreover, the linear independence of $x_1, x_2, \ldots, x_r$ immediately implies that $\beta^1, \beta^2, \ldots, \beta^r$ are uniquely determined by x. These real numbers are the components of x with respect to $x_1, x_2, \ldots, x_r$.

Thus we are justified in calling $x_1, x_2, \ldots, x_r$ a __basis__ for U. This completes the proof.

THEOREM 7.4.   Let U be a subspace of R^n. Then there exists another subspace W of R^n such that

(i)   If $x \in R^n$ there are uniquely determined vectors $u \in U$, $w \in W$ such that

$$x = u + w,$$

(ii) $\dim U + \dim W = n = \dim R^n$.

REMARK.   We say that R^n is the __direct sum__ of U and W and we write

$$R^n = U \oplus W.$$

We say that subspaces U and W are __complementary__ in R^n.

PROOF OF THEOREM 7.4.   Let us assume that $\dim U = r$ and let $x_1, x_2, \ldots, x_r$ be r linearly independent vectors in U. If $r = n$ then $x_1, x_2, \ldots, x_r$ must be a basis for R^n and we have $U = R^n$.   So we assume $r < n$.   Then $x_1, x_2, \ldots, x_r$ do not constitute a basis for R^n and there is some vector, which we shall call x_{r+1}, such that x_{r+1} is not a linear combination of $x_1, x_2, \ldots, x_r$. We claim then that $x_1, x_2, \ldots, x_r$, x_{r+1} are linearly independent. If this were not so there would be real numbers $\alpha^1, \alpha^2, \ldots, \alpha^r, \alpha^{r+1}$ such that

$$\alpha^1 x_1 + \alpha^2 x_2 + \ldots + \alpha^r x_r + \alpha^{r+1} x_{r+1} = 0.$$

As in the proof of Theorem 7.3, $\alpha^{r+1} \neq 0$ and we have

$$x_{r+1} = -\frac{\alpha^1}{\alpha^{r+1}} x_1 - \frac{\alpha^2}{\alpha^{r+1}} x_2 - \ldots - \frac{\alpha^r}{\alpha^{r+1}} x_r,$$

contradicting our assumption on x_{r+1}. Hence $x_1, x_2, \ldots, x_r$, x_{r+1} are linearly independent. We continue in this manner, obtaining vectors x_{r+j} such that $x_1, x_2, \ldots, x_r, x_{r+1}, \ldots, x_{r+j}$ are linearly independent. When $r + j = n$ we can proceed no further by definition of $\dim R^n = n$. It is clear then that $x_1, x_2, \ldots, x_n$ form a basis for R^n.

Let

$$W = \{x \in R^n \mid x = \alpha^{r+1} x_{r+1} + \ldots + \alpha^n x_n\}.$$

Since $x_1, x_2, \ldots, x_n$ form a basis for R^n, if $x \in R^n$ we have

$$x = \alpha^1 x_1 + \alpha^2 x_2 + \ldots + \alpha^n x_n$$

for some real numbers $\alpha^1, \alpha^2, \ldots, \alpha^n$. Putting

$$u = \alpha^1 x_1 + \alpha^2 x_2 + \ldots + \alpha^r x_r$$

$$w = \alpha^{r+1} x_{r+1} + \alpha^{r+2} x_{r+2} + \ldots + \alpha^n x_n$$

we have $u \in U$, $w \in W$, so $x = u + w$ as described in (i)

above. Now if we also have $x = \hat{u} + \hat{w}$ with $\hat{u} \in U$, $\hat{w} \in W$,

then

$$\hat{u} = \beta^1 x_1 + \beta^2 x_2 + \ldots + \beta^r x_r$$

since $x_1, x_2, \ldots, x_r$ is a basis for U and

$$\hat{w} = \beta^{r+1} x_{r+1} + \beta^{r+2} x_{r+2} + \ldots + \beta^n x_n$$

by definition of W. Then

$$0 = u - u = (\alpha^1 - \beta^1)x_1 + \ldots + (\alpha^r - \beta^r)x_r + (\alpha^{r+1} - \beta^{r+1})x_{r+1}$$

$$+ \ldots + (\alpha^n - \beta^n)x_n$$

and, since $x_1, x_2, \ldots, x_n$ are linearly independent, the

coefficients $\alpha^i - \beta^i = 0$ so that $\alpha^i = \beta^i$ and

$$u = \hat{u}, \quad w = \hat{w},$$

thus proving the uniqueness claimed in (i).

Since $x^{r+1}, x^{r+2}, \ldots, x^n$ all lie in W and are linearly

independent, $\dim W \geq n - r$. If $\dim W = q > n - r$ then there

are vectors $y_1, y_2, \ldots, y_q$, linearly independent and lying in

W. But then $x^1, x^2, \ldots, x^r, y^1, y^2, \ldots, y^q$ are linearly in-

dependent. For suppose we have

$$\alpha^1 x_1 + \ldots + \alpha^r x_r + \beta^1 y_1 + \ldots + \beta^q y_q = 0.$$

Putting

$$u = \alpha^1 x_1 + \ldots + \alpha^r x_r$$

$$w = \beta^1 y_1 + \ldots + \beta^q y_q$$

we have

$$0 = u + w, \quad u \in U, \quad w \in W.$$

But also

$$0 = 0 + 0$$

and $0 \in u$, $0 \in W$. By the uniqueness in (i) we then must

have $u = 0$, $w = 0$. But then, since $x^1, \ldots, x^r$ and

$y^1, \ldots, y^q$ are each sets of linearly independent vectors, we

must have $\alpha^1 = \ldots = \alpha^r = 0$, $\beta^1 = \ldots = \beta^q = 0$. Thus the equation $\alpha^1 x_1 + \ldots + \alpha^r x_r + \beta^1 y_1 + \ldots + \beta^q y_q = 0$ implies

all coefficients are zero and thus $x_1, \ldots, x_r, y_1, \ldots, y_q$ are

linearly independent. But then $\dim R^n \geq r + q > n$, a contra-

diction. We conclude therefore that $\dim W = n - r$ and (ii)

has been proved. With this the proof of Theorem 7.4 is

complete.

EXAMPLE. In R^3 let

$$U = \left\{ y = \begin{pmatrix} y^1 \\ y^2 \\ 0 \end{pmatrix} \in R^3 \,\middle|\, y^1, y^2 \text{ real} \right\},$$

let $\hat{x} = \begin{pmatrix} \hat{x}^1 \\ \hat{x}^2 \\ \hat{x}^3 \end{pmatrix}$ be a fixed vector in R^3 with $\hat{x}^3 \neq 0$ and put

$$W = \{ x \in R^3 \,|\, x = \alpha \hat{x}, \ \alpha \text{ real} \}.$$

Then $R^3 = U \oplus W$. To prove this we let $x = \begin{pmatrix} x^1 \\ x^2 \\ x^3 \end{pmatrix}$ be any

vector in R^3 and try to write

$$x = \begin{pmatrix} x^1 \\ x^2 \\ x^3 \end{pmatrix} = \begin{pmatrix} y^1 \\ y^2 \\ 0 \end{pmatrix} + \alpha \begin{pmatrix} \hat{x}^1 \\ \hat{x}^2 \\ \hat{x}^3 \end{pmatrix}.$$

The third component yields the equation $x^3 = \alpha \hat{x}^3$, whence

$\alpha = \dfrac{x^3}{\hat{x}^3}$. Then

$$y^1 = x^1 - (\frac{x^3}{\hat{x}^3})\hat{x}^1, \quad y^2 = x^2 - (\frac{x^3}{\hat{x}^3})\hat{x}^2 .$$

We see that y^1, y^2 and α exist and are uniquely determined

by x. Hence $R^3 = U \oplus W$ as claimed.

DEFINITION 7. 4. Let A be an $n \times n$ matrix. The real

number λ is an _eigenvalue_ of A if there exists at least one

non-zero vector $x_\lambda \in R^n$ such that

$$Ax_\lambda = \lambda x_\lambda .$$

Any non-zero vector $y \in R^n$ with the property

$$Ay = \lambda y$$

is called an _eigenvector_ of A corresponding to the eigenvalue

λ . (Note that if $\mu \neq 0$ is a real number then μy is also an

eigenvalue of A corresponding to λ .)

The condition $Ay = \lambda y$ is, of course, equivalent to

$$(\lambda I - A)y = 0 .$$

From Chapter 4 we know that this equation has a solution different from zero if and only if $\det(\lambda I - A) = 0$. Now the rules for computation of determinants show readily that

$$\det(\lambda I - A) = p(\lambda)$$

where

$$p(\lambda) = \lambda^n + a_1 \lambda^{n-1} + a_2 \lambda^{n-2} + \ldots + a_n$$

is a polynomial of degree n in λ called the <u>characteristic polynomial</u> of the matrix A. We see therefore that the eigenvalues of A are precisely the roots of the polynomial equation

$$p(\lambda) = 0.$$

The polynomial $p(\lambda)$ has a factorization

$$p(\lambda) = (\lambda - \lambda_1)(\lambda - \lambda_2) \ldots (\lambda - \lambda_n)$$

where $\lambda_1, \lambda_2, \ldots, \lambda_n$ are the roots of $p(\lambda) = 0$. If A has real entries the coefficients of the polynomial $p(\lambda)$, i. e., $a_1, a_2, \ldots, a_n$, will all be real numbers. Even so, some of the roots λ_i may be complex numbers and some numbers may be repeated in the sequence $\lambda_1, \lambda_2, \ldots, \lambda_n$. All of these eigenvalues are important but we will be primarily interested in the case where all of the λ_i are real numbers and are distinct, i. e., $\lambda_i \neq \lambda_j$ for $i \neq j$. For this case we have

THEOREM 7. 5. Let the $n \times n$ matrix A have n real distinct eigenvalues $\lambda_1, \lambda_2, \ldots, \lambda_n$ and let $x_1, x_2, \ldots, x_n$ be eigenvalues of A corresponding, respectively, to these eigenvalues. Then $x_1, x_2, \ldots, x_n$ are linearly independent and hence form a basis for R^n.

Proof. We will first show that x_1, x_2 are linearly independent. Let α^1 and α^2 be real numbers such that

$$\alpha^1 x_1 + \alpha^2 x_2 = 0.$$

Then

$$0 = A(\alpha^1 x_1 + \alpha^2 x_2) = \alpha^1 A(x_1) + \alpha^2 A(x_2)$$

$$= \alpha^1 \lambda_1 x_1 + \alpha^2 \lambda_2 x_2.$$

Since $\lambda_i \neq \lambda_j$, $i \neq j$, at most one of the n eigenvalues is equal to zero. Without loss of generality assume that if there is a zero eigenvalue it is $\lambda_1 = 0$. Then $\lambda_2 \neq 0$ and we can divide the last equation above by λ_2 to obtain

$$\alpha^1 \frac{\lambda_1}{\lambda_2} x_1 + \alpha^2 x_2 = 0.$$

From this we subtract the equation $\alpha^1 x_1 + \alpha^2 x_2 = 0$ obtaining

$$\alpha^1 (\frac{\lambda_1}{\lambda_2} - 1) x_1 = 0.$$

Now x_1 is not the zero vector and, since $\lambda_1 \neq \lambda_2$, $\frac{\lambda_1}{\lambda_2} \neq 1$.
Hence $\alpha^1 = 0$. But then $\alpha^2 x_2 = 0$ and, since x_2 is not
the zero vector, $\alpha^2 = 0$ also. We conclude that the equation
$\alpha^1 x_1 + \alpha^2 x_2 = 0$ implies $\alpha^1 = \alpha^2 = 0$ and hence x_1, x_2 are
linearly independent.

Now suppose that $2 \leq m < n$ and the vectors x_1,
$x_2, \ldots, x_m$ have been shown to be linearly independent. Let
$\alpha^1, \alpha^2, \ldots, \alpha^{m+1}$ be real numbers such that

$$\alpha^1 x_1 + \alpha^2 x_2 + \ldots + \alpha^m x_m + \alpha^{m+1} x_{m+1} = 0.$$

Applying A to both sides we obtain

$$\alpha^1 \lambda_1 x_1 + \alpha^2 \lambda_2 x_2 + \ldots + \alpha^m \lambda_m x_m + \alpha^{m+1} \lambda_{m+1} x_{m+1} = 0.$$

By assumption $\lambda_{m+1} \neq 0$. We divide the last equation by

λ_{m+1} and subtract the first equation to obtain

$$\alpha^1(\frac{\lambda_1}{\lambda_{m+1}} - 1)x_1 + \alpha^2(\frac{\lambda_2}{\lambda_{m+1}} - 1)x_2 + \ldots + \alpha^m(\frac{\lambda_m}{\lambda_{m+1}} - 1)x_m = 0 .$$

Since $x_1, x_2, \ldots, x_m$ are linearly independent the coefficients

$\alpha^i(\frac{\lambda_i}{\lambda_{m+1}} - 1) = 0$, $i = 1, 2, \ldots, m$. But $\lambda_i \neq \lambda_{m+1}$, so we

conclude $\alpha^i = 0$, $i = 1, 2, \ldots, m$. Then we have

$$\alpha^{m+1} x_{m+1} = 0$$

and, since $x_{m+1} \neq 0$, we have $\alpha^{m+1} = 0$ and hence $\alpha^i = 0$,

$i = 1, 2, \ldots, m + 1$. We conclude that $x_1, x_2, \ldots, x_{m+1}$ are

linearly independent. Continuing in this manner we will

finally have $x_1, x_2, \ldots, x_n$ linearly independent and the

proof is complete.

EXAMPLE. Let A be the 3×3 matrix

$$A = \begin{pmatrix} 0 & 1 & 0 \\ 0 & 0 & 1 \\ 6 & -11 & 6 \end{pmatrix} .$$

The characteristic polynomial is

$$p(\lambda) = \det \begin{pmatrix} \lambda & -1 & 0 \\ 0 & \lambda & -1 \\ -6 & 11 & \lambda-6 \end{pmatrix}$$

$$= \lambda^3 - 6\lambda^2 + 11\lambda - 6 = (\lambda - 1)(\lambda - 2)(\lambda - 3)$$

and thus A has eigenvalues $\lambda_1 = 1$, $\lambda_2 = 2$, $\lambda_3 = 3$. Let

$$y_\lambda = \begin{pmatrix} y_\lambda^1 \\ y_\lambda^2 \\ y_\lambda^3 \end{pmatrix}$$ be an eigenvalue corresponding to the eigenvalue

λ. Then

$$0 = (\lambda I - A)y_\lambda = \begin{pmatrix} \lambda & -1 & 0 \\ 0 & \lambda & -1 \\ -6 & 11 & \lambda-6 \end{pmatrix} \begin{pmatrix} y_\lambda^1 \\ y_\lambda^2 \\ y_\lambda^3 \end{pmatrix}.$$

The first two equations in $y_\lambda^1, y_\lambda^2, y_\lambda^3$ are

$$\lambda y_\lambda^1 - y_\lambda^2 = 0,$$

$$\lambda y_\lambda^2 - y_\lambda^3 = 0.$$

Take $y^3 = \lambda^2$ and we have $y_\lambda^2 = \lambda$, $y_\lambda^1 = 1$. Substituting these values in the third equation

$$-6y_\lambda^1 + 11y_\lambda^2 + (\lambda-6)y_\lambda^3 = -6(1) + 11(\lambda) + (\lambda-6)\lambda^2$$

$$= \lambda^3 - 6\lambda^2 + 11\lambda - 6 = 0 \quad \text{if} \quad \lambda = 1, 2, \text{ or } 3 .$$

Thus the vectors

$$y_1 = \begin{pmatrix} 1 \\ 1 \\ 1 \end{pmatrix}, \qquad y_2 = \begin{pmatrix} 1 \\ 2 \\ 4 \end{pmatrix}, \qquad y_3 = \begin{pmatrix} 1 \\ 3 \\ 9 \end{pmatrix}$$

are eigenvalues of A corresponding to the eigenvalues $\lambda_1 = 1$, $\lambda_2 = 2$, $\lambda_3 = 3$, respectively. Since

$$\det \begin{pmatrix} 1 & 1 & 1 \\ 1 & 2 & 3 \\ 1 & 4 & 9 \end{pmatrix} = 2$$

the vectors y_1, y_2, y_3 form a basis for R^3.

DEFINITION 7.5. Let $x_1, x_2, \ldots, x_r$ be vectors in R^n such that

$$x_i' x_j = \begin{cases} 1 & \text{if} \quad i = j \\ \\ 0 & \text{if} \quad i \neq j. \end{cases}$$

Then $x_1, x_2, \ldots, x_r$ are said to be an <u>orthonormal set</u> in R^n.

THEOREM 7.6. An orthonormal system of vectors in R^n is a linearly independent set. Thus an orthonormal system consisting of n vectors in R^n is a basis for R^n. Moreover, if $x_1, x_2, \ldots, x_n$ are orthonormal in R^n we have, for each $x \in R^n$,

$$x = (x_1' x)x_1 + (x_2' x)x_2 + \ldots + (x_n' x)x_n.$$

<u>Proof.</u> Suppose $\alpha^1, \alpha^2, \ldots, \alpha^r$ are real numbers such that

$$\alpha^1 x_1 + \alpha^2 x_2 + \ldots + \alpha^r x_r = 0.$$

Then for each x_i, $i = 1, 2, \ldots, r$, Definition 7.5 shows that

$$0 = x_i'(\alpha^1 x_1 + \alpha^2 x_2 + \ldots + \alpha^r x_r)$$

$$= \alpha^1 (x_i' x_1) + \alpha^2 (x_i' x_2) + \ldots + \alpha^r (x_i' x_r)$$

$$= \alpha^i.$$

Thus $\alpha^1 = \alpha^2 = \ldots = \alpha^r = 0$ and we conclude that $x_1, x_2, \ldots, x_r$ are linearly independent.

Now suppose $x_1, x_2, \ldots, x_n$ is an orthonormal set which is also a basis for R^n. (We will call such a set an <u>orthonormal basis.</u>) If $x \in R^n$ there are real numbers $\alpha^1, \alpha^2, \ldots, \alpha^n$ such that

$$x = \alpha^1 x_1 + \alpha^2 x_2 + \ldots + \alpha^n x_n.$$

Then for $i = 1, 2, \ldots, n$

$$x_i' x = x_i' (\alpha^1 x_1 + \alpha^2 x_2 + \ldots + \alpha^n x_n)$$

$$= \alpha^1 (x_i' x_1) + \alpha^2 (x_i' x_2) + \ldots + \alpha^n (x_i' x_n) = \alpha^i$$

and the proof is complete.

For general matrices we have shown only that eigenvalues $x_1, x_2, \ldots, x_n$ form a basis for R^n if the corresponding eigenvalues $\lambda_1, \lambda_2, \ldots, \lambda_n$ are distinct. For <u>symmetric</u> matrices we can obtain a stronger result. We begin by noting a very simple result: if A is a symmetric matrix and λ_1, λ_2 are distinct eigenvalues of A with corresponding eigenvectors x_1, x_2, then $x_1' x_2 = 0$. For

$$x_1' A x_2 = x_2' A' x_1 = x_2' A x_1$$

and therefore

$$0 = x_1' A x_2 - x_2' A x_1 = x_1' (\lambda_2 x_2) - x_2' (\lambda_1 x_1)$$

$$= (\lambda_2 - \lambda_1) x_1' x_2$$

and since $\lambda_2 - \lambda_1 \neq 0$, $x_1' x_2 = 0$. From this it is clear that if a symmetric $n \times n$ matrix A has n <u>distinct</u> eigenvalues

$\lambda_1, \lambda_2, \ldots, \lambda_n$, corresponding eigenvectors $x_1, x_2, \ldots, x_n$ will form an orthogonal basis for R^n. The next theorems show that this remains true even if the symmetric matrix A does not have n distinct eigenvalues.

THEOREM 7. 7. Let A be an n × n symmetric matrix. Let S be the subset of R^n defined by

$$S = \{y \in R^n \mid \|y\|_e = 1\} .$$

Let us define a continuous function $f : R^n \to R^1$ by

$$f(y) = y' Ay .$$

Since S is compact f achieves a maximum value at some point $y_1 \in S$ and we put $\lambda_1 = f(y_1)$. Then λ_1 is an eigenvalue of A and y_1 is a corresponding eigenvector.

Proof. Let $Ay_1 = z_1$ and then put

$$w_1 = z_1 - (y_1' z_1)y_1 .$$

Now w_1 is orthogonal to y_1, for

$$y_1'w_1 = y_1'(z_1 - (y_1'z_1)y_1)$$

$$= (y_1'z_1) - (y_1'z_1)(y_1'y_1) = 0$$

since $y_1'y_1 = \|y_1\|_e^2 = 1$. Let

$$y(\mu) = \frac{y_1 + \mu w_1}{\sqrt{1 + \mu^2 \|w_1\|_e^2}} \quad .$$

Now

$$\|y_1 + \mu w_1\|_e = \sqrt{(y_1 + \mu w_1)'(y_1 + \mu w_1)}$$

$$= \sqrt{y_1'y_1 + 2\mu y_1'w_1 + \mu^2 w_1'w_1} = \sqrt{1 + \mu^2 \|w_1\|_e^2} \quad .$$

We conclude therefore that $\|y(\mu)\|_e = 1$ so that $y(\mu) \in S$ for all μ such that $1 + \mu^2 \|w_1\|_e^2 \neq 0$, which includes all μ in some neighborhood of 0. Then we define a function $g : R^1 \to R^1$ by

$$g(\mu) = f(y(\mu)).$$

Since $f(y(\mu)) \leq f(y_1)$ for all μ, $g(\mu)$ has a maximum at $\mu = 0$. But

$$\frac{dq}{d\mu} = \frac{\partial f}{\partial y}(y(\mu))\frac{dy}{d\mu}$$

$$= 2y(\mu)'A\left(\frac{w_1}{1+\mu^2\|w_1\|_e^2} - \frac{\mu\|w_1\|_e^2 y(\mu)}{(1+\mu^2\|w_1\|_e^2)^{3/2}}\right)$$

which shows that

$$\frac{dq}{d\mu}(0) = 2y_1'Aw_1.$$

Since g has a maximum at $\mu = 0$ we must have $y_1'Aw_1 = z_1'w_1 = w_1'z_1 = 0$. But

$$w_1'z_1 = w_1'(w_1 - (y_1'z_1)y_1) = \|w_1\|_e^2$$

since $w_1'y_1 = y_1'w_1 = 0$. Thus $\|w_1\|_e^2 = 0$ and we have

$$Ay_1 = z_1 = (y_1'z_1)y_1$$

and thus

$$\lambda_1 = y_1'Ay_1 = y_1'(y_1'z_1)y_1 = (y_1'z_1)$$

and it follows that $Ay_1 = \lambda_1 y_1$, which proves the theorem.

It is clear that if λ is any other eigenvalue of A then $\lambda \leq \lambda_1$. For if y is an eigenvector of A corresponding to λ , we may assume $\|y\|_e = 1$ and then

$$\lambda = \lambda y'y = (Ay)'y = y'A'y = y'Ay \le y_1' Ay_1 = \lambda_1 .$$

Thus λ_1 is the largest eigenvalue of A. In the same way it can be shown that if ν is the minimum value of $f(y)$ on S then ν is the least eigenvalue of A.

LEMMA 7.1. (Gram Schmidt Process). Let V be an r-dimensional subspace of R^n, $r \le n$. Then V has an orthonormal basis.

Proof. If $r = 1$ there is nothing to prove. So assume $r \ge 2$. Let $x_1, x_2, \ldots, x_r$ be a basis for V. Clearly no $x_i = 0$. Thus we may put

$$y_1 = \frac{x_1}{\|x_1\|_e}$$

and we have $\|y_1\|_e = 1$. Next put

$$\hat{y}_2 = x_2 - (y_1' x_2) y_1 .$$

Then $y_1' \hat{y}_2 = 0$. Moreover, $\hat{y}_2 \ne 0$, otherwise we would have

$$0 = x_2 - \frac{(y_1' x_2)}{\|x_1\|_e} x_1$$

and x_1, x_2 would be linearly dependent, contrary to our assumption. Setting $y_2 = \dfrac{\hat{y}_2}{\|\hat{y}_2\|_e}$ we have

$$\|y_2\|_e^2 = 1, \qquad y_1' y_2 = 0$$

and Theorem 7.6 applies to show y_1, y_2 linear independent.

If $r = 2$ we are done. Otherwise assume we have $y_1, y_2, \dots, y_s$, $2 \leq s \leq r-1$ such that y_i, $i = 1, 2, \dots, s$ is a linear combination of $x_1, x_2, \dots, x_i$ and $y_1, y_2, \dots, y_s$ is an orthonormal set in V. Define $\hat{y}_{s+1}$ by

$$\hat{y}_{s+1} = x_{s+1} - (y_1' x_{s+1}) y_1 - (y_2' x_{s+1}) y_2 - \dots - (y_s' x_{s+1}) y_s .$$

Because $y_1, y_2, \dots, y_s$ form an orthonormal set we easily see that $y_i' y_{s+1} = 0$, $i = 1, 2, \dots, s$. Also, $\hat{y}_{s+1} \neq 0$, otherwise

$$0 = x_{s+1} - (y_1' x_{s+1}) y_1 - (y_2' x_{s+1}) y_2 - \dots - (y_s' x_{s+1}) y_s$$

implies $x_1, x_2, \dots, x_s, x_{s+1}$ linearly dependent, since each y_i is a linear combination of $x_1, x_2, \dots, x_i$. Putting

$$y_{s+1} = \frac{\hat{y}_{s+1}}{\|\hat{y}_{s+1}\|_e}$$

we see that $y_1, y_2, \ldots, y_{s+1}$ form an orthonormal set.

We continue this process until we have an orthonormal set consisting of r vectors in V. Such a set forms an orthonormal basis for V and the proof is complete.

The main result of this chapter is

THEOREM 7.8. Let A be an $n \times n$ symmetric matrix. Then there is an orthonormal basis for R^n consisting of eigenvectors of A.

Proof. Let λ_1 be the largest eigenvalue of A as found in Theorem 7.7. Let

$$V_1 = \{x \in R^n \,|\, Ax = \lambda_1 x\}.$$

Without difficulty we verify that V_1 is a subspace of R^n. Let $\dim V_1 = r_1$ and let $y_{11}, y_{12}, \ldots, y_{1r_1}$ be an orthonormal basis for V_1. Each of these vectors is an eigenvector of A corresponding to the eigenvalue λ_1.

Let

$$M_1 = \{x \in R^n \mid y'x = 0 \quad \text{whenever} \quad y \in V_1\}.$$

Again we verify without difficulty that M_1 is a subspace of R^n. Moreover,

$$R^n = V_1 \oplus M_1.$$

For if $x \in R^n$ we can put

$$y = x - (y'_{11}x)y_{11} - \ldots - (y'_{1r_1}x)y_{1r_1} = x - w$$

so that $x = y + w$, $y \in M_1$, $w \in V_1$.

We now claim that $A : M_1 \rightarrow M_1$. For if $x \in M_1$ and $y \in V_1$, we have

$$y' Ax = (Ay)' x = \lambda_1 y'x = 0$$

which shows $Ax \in M_1$ as well as x.

We now treat A as a linear function $A : M^1 \rightarrow M^1$ and consider the same problem as in Theorem 7.7 but with R^n replaced by M_1. We let $f_1(y) = y' Ay$ be defined for just those $y \in M_1$ and we seek its maximum over vectors $y \in M_1$ with $\|y\|_e = 1$. We let y_2 be a solution of this problem and set $f_1(y_2) = \lambda_2$. Just as in Theorem 7.7 we verify that

λ_2 is an eigenvalue of A and y_2 an associated eigenvector.
We put

$$V_2 = \{y \in M_1 \mid Ay = \lambda_2 y_2\}$$

$$M_2 = \{x \in M_1 \mid y'x = 0, \ y' \in V_2\}.$$

One can now verify that $A : M_2 \to M_2$ and

$$R^n = V_1 \oplus V_2 \oplus M_2.$$

We continue in this manner. At the m-th stage we see
that $R^n = V_1 \oplus \ldots \oplus V_{m-1} \oplus M_{m-1}$ and $A : M_{m-1} \to M_{m-1}$.
We let $f_{m-1}(y) = y'Ay$ be defined for $y \in M_{m-1}$ and seek its
maximum over $y \in M_{m-1}$ with $\|y\|_e = 1$. This maximum is
achieved at y_m and $f(y_m) = \lambda_m$ is an eigenvalue of A.
Eventually for some m we will have $\dim V_1 + \ldots + \dim V_m = n$
and then

$$R^n = V_1 \oplus V_2 \oplus \ldots \oplus V_m.$$

We let $\dim V_i = r_i$, $i = 1, 2, \ldots, m$, and take $y_{i1}, y_{i2}, \ldots, y_{ir_i}$
to be an orthonormal basis for V_i. One can then verify easily
that $y_{11}, \ldots, y_{1r_1}, y_{21}, \ldots, y_{2r_2}, \ldots, y_{m1}, \ldots, y_{mr_m}$ is an
orthonormal basis for R^n and each of these vectors is an

eigenvector of A. In fact $y_{i1}, \ldots, y_{ir_i}$ are eigenvectors of A corresponding to the eigenvalue λ_i. With this our proof is complete.

A matrix A is said to be <u>diagonal</u> if $A = (a_i^j)$ and $a_i^j = 0$ for $j \neq i$. In this case

$$A = \begin{pmatrix} a_1^1 & 0 & \cdots & 0 \\ 0 & a_2^2 & \cdots & 0 \\ \vdots & \vdots & \ddots & \vdots \\ 0 & 0 & \cdots & a_n^n \end{pmatrix} \equiv \mathrm{diag}\,(a_1^1, a_2^2, \ldots, a_n^n) \ .$$

Then we easily verify that each a_i^i, $i = 1, 2, \ldots, n$, is an eigenvalue of A and the vector e_i defined by $e_i^i = 1$, $e_i^j = 0$, $j \neq i$, is an eigenvector of A corresponding to the eigenvalue a_i^i. It is a noteworthy fact that if R^n has a basis consisting of eigenvectors of A then A can be transformed in a natural way so that it becomes a diagonal matrix.

Let $y = f(x)$ be a linear function, $f : R^n \to R^n$. In matrix notation

$$y = Ax.$$

In R^n we make a change of variables or transformation. We put

$$x = Pz$$

$$y = Pw$$

where P is a non-singular $n \times n$ matrix. Then the equation $y = Ax$ becomes $Pw = APz$ or

$$w = P^{-1} APz = Bz.$$

The formulas $y = Ax$ and $w = Bz$ represent the same linear function f but they describe f in terms of different coordinate systems or bases for R^n.

DEFINITION 7.6. If A and B are $n \times n$ matrices and there is a non-singular $n \times n$ matrix P such that

$$B = P^{-1} AP$$

then A and B are said to be <u>similar</u>.

THEOREM 7.9. Let A be an $n \times n$ matrix and let $y_1, y_2, \ldots,$ y_n be a basis for R^n consisting of eigenvectors of A corresponding to eigenvalues $\lambda_1, \lambda_2, \ldots, \lambda_n$, respectively. Then there is a non-singular $n \times n$ matrix P such that

(i) $P^{-1}AP = \Lambda = \text{diag}(\lambda_1, \lambda_2, \ldots, \lambda_n)$;

(ii) $Pe_i = y_i$, $i = 1, 2, \ldots, n$, where

$$e^i_i = 1, \quad e^j_i = 0, \quad j \neq i.$$

(iii)   If A is symmetric we can choose P so that

$$P^{-1} = P'.$$

REMARKS.   We summarize (i) by saying that P <u>diagonalizes</u> A.   If, as in (iii), $P^{-1} = P'$ we say that P is an <u>orthogonal</u> matrix.

PROOF OF THEOREM 7.9.   Let

$$y_i = \begin{pmatrix} y^1_i \\ y^2_i \\ \vdots \\ y^n_i \end{pmatrix}, \; i = 1,2,\ldots,n, \quad P = \begin{pmatrix} y^1_1 & y^1_2 & \cdots & y^1_n \\ y^2_1 & y^2_2 & \cdots & y^2_n \\ \vdots & \vdots & & \vdots \\ y^n_1 & y^n_2 & \cdots & y^n_n \end{pmatrix}.$$

We will write $P = (y_1, y_2, \ldots, y_n)$.   Since $P^{-1}P = I$, if we put

$$P^{-1} = \begin{pmatrix} z'_1 \\ z'_2 \\ \vdots \\ z'_n \end{pmatrix}$$

then $z_i'y_j$ is the $\binom{i}{j}$-th entry of the identity matrix so that

$$z_i'y_j = \begin{cases} 1, & i = j \\ 0, & i \neq j \end{cases}.$$ Now we compute

$$P^{-1}AP = P^{-1}(y_1, y_2, \ldots, y_n) = P^{-1}(Ay_1, Ay_2, \ldots, Ay_n)$$

$$= P^{-1}(\lambda_1 y_1, \lambda_2 y_2, \ldots, \lambda_n y_n) = \begin{pmatrix} z_1' \\ z_2' \\ \vdots \\ z_n' \end{pmatrix} (\lambda_1 y_1, \lambda_2 y_2, \ldots, \lambda_n y_n)$$

$$= \text{diag}(\lambda_1, \lambda_2, \ldots, \lambda_n).$$

Thus (i) has been proved. To show (ii) we compute

$$Pe_i = \sum_{j=1}^{n} e_i^j y_j = y_i.$$

If A is symmetric we can take $y_1, y_2, \ldots, y_n$ to be an ortho-normal basis for R^n. Then

$$P'P = \begin{pmatrix} y_1' \\ y_2' \\ \vdots \\ y_n' \end{pmatrix} (y_1, y_2, \ldots, y_n) = I$$

which proves (iii). Thus the proof of the theorem is complete.

EXERCISES.   CHAPTER 7

1.   Let V be an r-dimensional subspace of R^n, $r < n$, and let $y \in R^n$, $y \notin V$. Show that there is exactly one point $x^* \in V$ such that

$$\|y - x^*\|_e \leq \|y - x\|_e, \qquad x \in V.$$

Prove that $(y - x^*)' x^* = 0$. If $x_1, x_2, \ldots, x_r$ is an orthonormal basis for V, show that $x^* = \sum_{i=1}^{r} (y' x_i) x_i$.

2.   Let V be as in Problem 1. Define a linear transformation $P : R^n \to R^n$ by

$$Py = \begin{cases} y & \text{if } y \in V \\ x^* & \text{if } y \notin V. \end{cases}$$

Let $Q = I - P$. Show that $P^2 = PP = P$, $Q^2 = Q$, $PQ = QP = 0$. Show that P carries R^n into V and Q carries R^n into the subspace

$$V^\perp = \{y \in R^n \mid y' x = 0 \quad \text{for all} \quad x \in V\}.$$

Show that $R^n = V \oplus V^\perp$. (P and Q are the <u>orthogonal projections</u> from R^n onto V and $V^\perp$, respectively.)

3. As in Theorem 7.9 , let P be an orthogonal matrix, i. e. $P^{-1} = P'$. Show that the columns of P and also the rows of P are orthonormal bases for R^n. Show that for $x, y \in R^n$, $x'y = (Px)'Py$. Show that every orthogonal 2×2 matrix has the form

$$P = \begin{pmatrix} \cos \theta & \sin \theta \\ -\sin \theta & \cos \theta \end{pmatrix}$$

for some $\theta \in [0, 2\pi)$. Let $1 \le k < \ell \le n$ and let $P^k_\ell(\theta)$ be an $n \times n$ matrix with entries P^i_j given by

$$P^i_j = 1 \text{ if } i = j \text{ and } i \ne k \text{ or } \ell$$

$$P^k_k = p^\ell_\ell = \cos \theta$$

$$p^k_\ell = \sin \theta$$

$$p^\ell_k = -\sin \theta$$

$$p^i_j = 0 \text{ otherwise.}$$

Show that $P^k_\ell(\theta)$ is an orthogonal matrix.

4. Let A be an $n \times n$ matrix and let P be an orthogonal matrix. Let

$$B = P^{-1}AP = P'AP.$$

Prove that if $A = (a_i^j)$, $B = (b_i^j)$, then

$$\sum_{i=1}^{n} \sum_{j=1}^{n} (a_i^j)^2 = \sum_{i=1}^{n} \sum_{j=1}^{n} (b_i^j)^2 \,.$$

Hint: Let $C = A'A = (C_i^i)$. Then

$$\sum_{i=1}^{n} \sum_{j=1}^{n} (a_i^j)^2 = \sum_{i=1}^{n} (C_i^i)^2 \,.$$

5. Let A be an $n \times n$ symmetric matrix and let S_A be the set of $n \times n$ matrices

$$S_A = \{B \,|\, B = P'AP, \quad P \text{ orthogonal}\} \,.$$

For each $B \in S_A$ define

$$f(B) = \sum_{i=1}^{n} (b_i^i)^2 \,.$$

Prove that if $f(B_1) \geq f(B)$ for all $B \in S_A$ then B_1 is diagonal. Hint: If B_1 is not diagonal there is a matrix $P_\ell^k(\theta)$ as described in Problem 3 such that $f(B_1) < f((P_\ell^k(\theta))' B_1 P_\ell^k(\theta))$.

6. Devise a numerical method for diagonalizing a symmetric matrix A using the results of Problem 5. Put

$$A_0 = A$$

$$A_{m+1} = (P_{\ell_m}^{k_m}(\theta_m))' A_m P_{\ell_m}^{k_m}(\theta_m)$$

where k_m, ℓ_m, θ_m are selected so that the largest off-diagonal element of A_m is reduced to zero in A_{m+1}.
(Look up Jacobi's method in a numerical analysis text.)

7.   Diagonalize the matrix

$$A = \begin{pmatrix} 4 & 1 & 2 \\ 1 & 6 & 1 \\ 2 & 1 & 8 \end{pmatrix}$$

using Jacobi's method.

CHAPTER 8. GRADIENT METHODS

Throughout this chapter we will assume $f : R^n \to R^1$ is continuously differentiable in some domain $D \subseteq R^n$ and that f assumes a local minimum value in D at a point $x^* \in D^o$. We will make other assumptions as needed.

In order to motivate the formula used in what is commonly called the gradient method we let λ be a positive number and consider

$$f(x - \lambda \frac{\partial f}{\partial x}(x)') = f(x) + \frac{\partial f}{\partial x}(x)(- \lambda \frac{\partial f}{\partial x}(x)') + \mathcal{O}(\lambda)$$

$$= f(x) - \lambda \left\| \frac{\partial f}{\partial x}(x) \right\|_e^2 + \mathcal{O}(\lambda).$$

If $\frac{\partial f}{\partial x}(x) \neq 0$ then for sufficiently small $\lambda > 0$ we clearly have

$$f(x - \lambda \frac{\partial f}{\partial x}(x)') < f(x).$$

Thus if we are searching for a minimum of f the point $x - \lambda \frac{\partial f}{\partial x}(x)'$ is an improvement over the point x if $\frac{\partial f}{\partial x}(x) \neq 0$ and λ is sufficiently small. By repeated construction of new points in this manner we may hope to approach x^*, a local minimum of f in D. The gradient method consists in the construction of a sequence $\{x_k\}$ of points in R^n by means of the recursion equation

$$x_{k+1} = x_k - \lambda \frac{\partial f}{\partial x}(x_k)', \qquad k = 0, 1, 2, 3, \ldots,$$

where x_0 is some given initial point. The first part of this chapter will be devoted to discovering whether or not

$$\lim_{k \to \infty} x_k = x^*$$

if we choose x_0 close to x^*.

First of all, continuous differentiability of f is not enough to ensure that the gradient method has this desirable property. Consider the example

$$f(x) = x^{4/3}, \qquad -\infty < x < \infty,$$

which is continuously differentiable for $-\infty < x < \infty$ and has

only one local minimum, namely $x^* = 0$. For this function

the gradient method involves the recursion equation

$$x_{k+1} = x_k - \frac{4}{3} \lambda x_k^{1/3} .$$

We let $I(\lambda)$ be the interval

$$I(\lambda) = [-(\frac{2}{3}\lambda)^{3/2}, (\frac{2}{3}\lambda)^{3/2}]$$

which has positive length for any $\lambda > 0$. Let $x_0 \in I(\lambda)$; for

convenience take $x_0 > 0$ as the analysis for $x_0 < 0$ is simi-

lar. Then

$$x_1 = x_0 - \frac{4}{3}\lambda x_0^{1/3} .$$

If $0 < x_0 < (\frac{2}{3}\lambda)^{3/2}$ then $0 < x_0^{2/3} < \frac{2}{3}\lambda$ and

$$x_1 < x_0 - 2(x_0^{2/3})x_0^{1/3} = -x_0$$

and we see that x_1 lies farther from $x^* = 0$ than x_0 does.

The same argument shows that $|x_k|$ increases with k as

long as $x_k \in I(\lambda)$, $x_k \neq 0$. This means that we cannot have

$$\lim_{k \to \infty} x_k = 0$$

unless there is some integer k_0 such that $x_{k_0} = 0$, which
will be true only if there is some special relationship between
x_0 and λ. In particular it is definitely not true that
$\lim_{k \to \infty} x_k = 0$ if we choose x_0 close enough to 0 and $\lambda > 0$
sufficiently small.

Now let us consider a different case, namely

$$f(x) = cx^2, \quad c > 0, \quad -\infty < x < \infty.$$

which again has only one local minimum at $x^* = 0$. Here the
gradient method gives

$$x_{k+1} = x_k - 2\lambda c x_k = (1 - 2\lambda c)x_k$$

and we see that for any x_0

$$x_k = (1 - 2\lambda c)^k x_0, \quad k = 0, 1, 2, \ldots,$$

and thus $\lim_{k \to \infty} x_k = 0$ if and only if

$$|1 - 2\lambda c| < 1$$

which is equivalent to

$$0 < \lambda < \frac{1}{c}.$$

Thus, in this case, the gradient method works if $0 < \lambda < \frac{1}{c}$

and fails if $\lambda \geq \frac{1}{c}$. The "best" value of λ is $\lambda = \frac{1}{2c}$ in

which case the minimum $x^* = 0$ is reached in one step. We

ask the reader to note that $2c$ is the Hessian (i. e. , the

second derivative in this case) of the function f.

Our first task will be to investigate <u>local convergence</u>

of the gradient method, i. e. , to answer the question: If f

has a local minimum at x^* and if x_0 is chosen sufficiently

close to x^* and we generate a sequence $\{x_k\}$ by $x_{k+1} =$

$x_k - \lambda \frac{\partial f}{\partial x}(x_k)'$, will it be true that $\lim_{k \to \infty} x_k = x^*$? Before

proving the theorem which answers this question we need

some auxiliary results.

THEOREM 8.1 (Contraction fixed point theorem). Let S be

a closed subset of R^n and let $g : R^n \to R^n$ have the property

that $g : S \to S$, i. e. , if $x \in S$, $g(x) \in S$. Suppose in addition

that g is a contraction on S, i. e. , there is a number γ,

$0 \leq \gamma < 1$, such that

$$\|g(x) - g(y)\| \leq \gamma \|x - y\|, \quad x, y \in S.$$

Then there is exactly one point $\hat{x} \in S$ such that $g(\hat{x}) = \hat{x}$.

Moreover, if x_0 is any point in S and we define a sequence $\{x_k\}$ by

$$x_{k+1} = g(x_k), \qquad k = 0, 1, 2, \ldots$$

then

$$\lim_{k \to \infty} x_k = \hat{x}.$$

Proof. Let $x_0 \in S$ and let a sequence $\{x_k\}$ be constructed as indicated in the theorem statement. Since $x_0 \in S$ and $g : S \to S$ it is clear that $x_k \in S$ for all k. Now estimate

$$\|x_{k+1} - x_k\| = \|g(x_k) - g(x_{k-1})\| \leq \gamma \|x_k - x_{k-1}\|$$

and deduce that

$$\|x_{k+1} - x_k\| \leq \gamma^k \|x_1 - x_0\|.$$

Let k, ℓ be positive integers, without loss of generality take $k > \ell$. Then

$$\|x_k - x_\ell\| \leq \|x_k - x_{k-1}\| + \|x_{k-1} - x_{k-2}\| + \ldots + \|x_{\ell+1} - x_\ell\|$$

$$\leq (\gamma^{k-1} + \gamma^{k-2} + \ldots + \gamma^\ell) \|x_1 - x_0\|$$

$$\leq (\sum_{j=\ell}^{\infty} \gamma^j) \|x_1 - x_0\| = \frac{\gamma^\ell}{1-\gamma} \|x_1 - x_0\| .$$

From this we see that, since $0 \leq \gamma < 1$, $\lim_{k, \ell \to \infty} \|x_k - x_\ell\| = 0$

and we conclude $\{x_k\}$ is a Cauchy sequence in S. Then $\{x_k\}$ has a limit $\hat{x}$, i.e.

$$\lim_{k \to \infty} x_k = \hat{x} .$$

Since S is closed, $\hat{x} \epsilon$ S. Now

$$x_{k+1} = g(x_k)$$

and g is clearly continuous. Taking limits on both sides

$$\hat{x} = \lim_{k \to \infty} x_{k+1} = \lim_{k \to \infty} g(x_k) = g(x)$$

and thus $\hat{x} = g(\hat{x})$, i. e., $\hat{x}$ is a <u>fixed point</u>.

If $\hat{x}$ and $\tilde{x}$ were both fixed points of g in S we would have

$$\|\hat{x} - \tilde{x}\| = \|g(\hat{x}) - g(\tilde{x})\| \leq \gamma \|\hat{x} - \tilde{x}\|$$

giving $(1 - \gamma) \| \hat{x} - \tilde{x} \| = 0$ so that $\| \hat{x} - \tilde{x} \| = 0$ and $x = \tilde{x}$.
This completes the proof.

THEOREM 8.2. Let $g : R^n \to R^n$ be continuously differenti-
able near a point $\hat{x} \in R^n$ such that $g(\hat{x}) = \hat{x}$. If $\| \frac{\partial g}{\partial x} (x) \| < 1$
there is an $\epsilon > 0$ such that if $\| \hat{x} - x_0 \| \leq \epsilon$ and $\{ x_k \}$ is
generated by $x_{k+1} = g(x_k)$, $k = 0, 1, 2, \ldots$, then

$$\lim_{k \to \infty} x_k = \hat{x}.$$

<u>Proof.</u> Since $\| \frac{\partial g}{\partial x} (\hat{x}) \| < 1$ we can find γ such that
$\| \frac{\partial g}{\partial x} (\hat{x}) \| < \gamma < 1$. Then, since $\frac{\partial g}{\partial x}$ is continuous near $\hat{x}$
there is an $\epsilon > 0$ such that $\| \frac{\partial g}{\partial x} (x) \| \leq \gamma$ when $\| \hat{x} - x \| \leq \epsilon$.
We define a closed set S by

$$S = \{ x \in R^n \,|\, \| \hat{x} - x \| \leq \epsilon \}.$$

Let x_1 and x_2 be any two points in S and define

$$x(\lambda) = \lambda x_2 + (1 - \lambda) x_1.$$

Then we compute

$$g(x_2) - g(x_1) = g(x(1)) - g(x(0))$$

$$= \int_0^1 \frac{d}{d\lambda} g(x(\lambda))d\lambda = \int_0^1 \frac{\partial g}{\partial x}(x(\lambda)) \frac{dx(\lambda)}{d\lambda} d\lambda$$

$$= \int_0^1 \frac{\partial g}{\partial x}(x(\lambda))(x_2 - x_1)d\lambda = (\int_0^1 \frac{\partial g}{\partial x}(x(\lambda))d\lambda)(x_2 - x_1).$$

Then, since $x(\lambda) \in S$, $0 \leq \lambda \leq 1$,

$$\|g(x_2) - g(x_1)\| \leq \|\int_0^1 \frac{\partial g}{\partial x}(x(\lambda))d\lambda\| \, \|x_2 - x_1\|$$

$$\leq (\int_0^1 \|\frac{\partial g}{\partial x}(x(\lambda))\| d\lambda) \|x_2 - x_1\| \leq \gamma \|x_2 - x_1\|$$

so that $\|g(x_2) - g(x_1)\| < \gamma \|x_2 - x_1\|$ whenever $x_1, x_2 \in S$.
If we can show $g : S \rightarrow S$ then g is a contraction on S. To
demonstrate this we let $x \in S$ and compute

$$\|\hat{x} - g(x)\| = \|g(\hat{x}) - g(x)\| \leq \gamma \|\hat{x} - x\| \leq \gamma\epsilon < \epsilon$$

which implies $g(x) \in S$.

   Once we know g is a contraction on S, Theorem 1
shows that $\hat{x}$ is the only fixed point of g in S and if
$x_0 \in S$ and $\{x_k\}$ is constructed as indicated, then

$$\lim_{k \to \infty} x_k = \hat{x}$$

and the proof is complete.

Note that in Theorems 8.1 and 8.2 we have not specified which norm we are dealing with. That is because these
theorems are valid no matter which norm is used. This can
lead to rather interesting situations as indicated in the following

EXAMPLE. Let $f : R^3 \to R^3$ be defined by $f(x) = Ax - b$,
where

$$
A = \begin{pmatrix} \frac{1}{8} & \frac{1}{27} & \frac{1}{64} \\ \frac{1}{4} & \frac{1}{9} & \frac{1}{16} \\ \frac{1}{2} & \frac{1}{3} & \frac{1}{4} \end{pmatrix} , \qquad b = \begin{pmatrix} 2 \\ 3 \\ 4 \end{pmatrix} .
$$

Seeking to solve $Ax = b$ we note that the solution $\hat{x}$ is a
fixed point of $g(x)$ where

$$
g(x) = x + Ax - b,
$$

that is, if

$$
\hat{x} = g(\hat{x}) = \hat{x} + A\hat{x} - b
$$

then clearly $A\hat{x} - b = 0$. Now $\dfrac{\partial g}{\partial x} \equiv A$ in R^n and thus, if we put $x_{k+1} = x_k + Ax_k - b$ we will have $\lim\limits_{k \to \infty} x_k = \hat{x}$ for

any $x_0 \in R^n$ provided $\left\| \dfrac{\partial g}{\partial x} \right\| = \|A\| < 1$.

Suppose we work with $\| \ \|_s$. Take $x = \begin{pmatrix} 1 \\ 1 \\ 1 \end{pmatrix}$. Then

$\|x\|_s = 1$. But $Ax = \begin{pmatrix} \dfrac{1}{8} + \dfrac{1}{27} + \dfrac{1}{64} \\[2mm] \dfrac{1}{4} + \dfrac{1}{9} + \dfrac{1}{16} \\[2mm] \dfrac{1}{2} + \dfrac{1}{3} + \dfrac{1}{4} \end{pmatrix}$ so that $\|Ax\|_s =$

$\dfrac{1}{2} + \dfrac{1}{3} + \dfrac{1}{4} = \dfrac{13}{12} > 1$, and we conclude $\|A\|_s > 1$. So if we

were to use only the $\| \ \|_s$ norm we might conclude that we

would not have $\lim\limits_{k \to \infty} x_k = \hat{x}$.

But now let us use $\| \ \|_t$. Put $x = \begin{pmatrix} x^1 \\ x^2 \\ x^3 \end{pmatrix}$, so that

$\|x\|_t = |x^1| + |x^2| + |x^3|$ and compute

$$Ax = \begin{pmatrix} \dfrac{1}{8}x^1 + \dfrac{1}{27}x^2 + \dfrac{1}{64}x^3 \\[2mm] \dfrac{1}{4}x^1 + \dfrac{1}{9}x^2 + \dfrac{1}{16}x^3 \\[2mm] \dfrac{1}{2}x^1 + \dfrac{1}{3}x^2 + \dfrac{1}{4}x^3 \end{pmatrix}$$

and we see that

$$\|Ax\|_t = |\frac{1}{8}x^1 + \frac{1}{27}x^2 + \frac{1}{64}x^3|$$

$$+ |\frac{1}{4}x^1 + \frac{1}{9}x^2 + \frac{1}{16}x^3| + |\frac{1}{2}x^1 + \frac{1}{3}x^2 + \frac{1}{4}x^3|$$

$$\leq \frac{7}{8}|x^1| + \frac{13}{27}|x^2| + \frac{21}{64}|x^3|$$

$$\leq \frac{7}{8}(|x^1| + |x^2| + |x^3|) = \frac{7}{8}\|x\|_t$$

so that $\|A\|_t = \frac{7}{8}$. Therefore Theorems 1 and 2 apply using the $\|\ \|_t$ norm and we do indeed have

$$\lim_{k \to \infty} x_k = \hat{x},$$

where $\hat{x}$ solves $Ax = b$, provided the x_k are generated as indicated, i. e.,

$$x_{k+1} = x_k + Ax_k - b.$$

The next theorem combines the results of Theorems 8.1 and 8.2 to answer the question of local convergence of the gradient method.

THEOREM 8.3. Let $f : R^n \to R^1$ be twice continuously dif-ferentiable near a point $x^* \in R^n$ where

(i) $\dfrac{\partial f}{\partial x}(x^*) = 0$;

(ii) $\dfrac{\partial^2 f}{\partial x^2}(x^*)$ is positive definite.

(Thus f has a local minimum at x^*.) Let λ_1 be the largest

eigenvalue of the symmetric matrix $\dfrac{\partial^2 f}{\partial x^2}(x^*)$. If λ is chosen

so that

$$0 < \lambda < \frac{2}{\lambda_1}$$

then there is an $\epsilon > 0$ such that if $\|x^* - x_0\| \le \epsilon$ and $\{x_k\}$

is generated by the gradient method

$$x_{k+1} = x_k - \lambda \frac{\partial f}{\partial x}(x_k)'$$

we will have

$$\lim_{k \to \infty} x_k = x^*.$$

Moreover, if the real number ρ satisfies

$$\max_{i = 1, 2, \ldots, n} \{(1 - \lambda \lambda_i)\} < \rho < 1$$

then for all sufficiently large k

$$\|x^* - x_{k+1}\|_e \leq \rho \|x^* - x_k\|_e.$$

<u>Proof.</u> We define a function $g : R^n \to R^n$ by

$$g(x) = x - \lambda \frac{\partial f}{\partial x}(x)'.$$

The gradient method is then just $x_{k+1} = g(x_k)$. Since f is twice continuously differentiable near x^*, g is continuously differentiable near x^*. From Theorem 8.2 we see that there is an $\epsilon > 0$ such that $\lim_{k \to \infty} x_k = x^*$ whenever $\|x^* - x_0\| \leq \epsilon$ provided that $\|\frac{\partial g}{\partial x}(x^*)\| \leq 1$. Now we compute

$$\frac{\partial g}{\partial x} = \frac{\partial}{\partial x}(x - \lambda \frac{\partial f'}{\partial x}) = I - \lambda \frac{\partial^2 f}{\partial x^2}$$

so that

$$\left\|\frac{\partial g}{\partial x}(x^*)\right\| = \left\|I - \lambda \frac{\partial^2 f}{\partial x^2}(x^*)\right\|.$$

Since $\frac{\partial^2 f}{\partial x^2}(x^*)$ is symmetric, Theorem 7.9 shows there is an orthogonal matrix P diagonalizing $\frac{\partial^2 f}{\partial x}(x^*)$. That is

$$P^{-1}\frac{\partial^2 f}{\partial x}(x^*)P = \text{diag}(\lambda_1, \lambda_2, \ldots, \lambda_n).$$

We see then that

$$P^{-1}(I - \lambda \frac{\partial^2 f}{\partial x^2}(x^*))P = \text{diag}(1-\lambda\lambda_1, 1-\lambda\lambda_2, \ldots, 1-\lambda\lambda_n).$$

We claim now that $\|\frac{\partial g}{\partial x}(x^*)\|_e = \|I - \lambda \frac{\partial^2 f}{\partial x^2}(x^*)\|_e$

$$= \max_{i=1,2,\ldots,n} |1 - \lambda\lambda_i|.$$

Let $y \in R^n$ and let $z = P^{-1}y = P'y$. Then

$$\|y\|_e^2 = y'y = (Pz)'Pz = z'P'Pz = z'z = \|z\|_e^2.$$

On the other hand, putting $A = I - \lambda \frac{\partial^2 f}{\partial x^2}(x^*)$,

$$\|Ay\|_e^2 = (Ay)'Ay = y'A'Ay = a'P'A'APz = z'P'A'PP'APz$$

$$= z'[\text{diag}((1-\lambda\lambda_1)^2, (1-\lambda\lambda_2)^2, \ldots, (1-\lambda\lambda_n)^2)]z.$$

If $z = \begin{pmatrix} z^1 \\ z^2 \\ \vdots \\ z^n \end{pmatrix}$ then

$$z'[\text{diag}((1-\lambda\lambda_1)^2, (1-\lambda\lambda_2)^2, \ldots, (1-\lambda\lambda_n)^2)]z$$

$$= \sum_{i=1}^{n}(1-\lambda\lambda_i)^2(z^i)^2 \leq \max_{i=1,2,\ldots,n}\{(1-\lambda\lambda_i)^2\}\sum_{i=1}^{n}(z^i)^2$$

$$= \max_{i=1,2,\ldots,n}\{(1-\lambda\lambda_i)^2\}\|z\|_e^2.$$

Thus we have

$$\|Ay\|_e^2 \le \max_{i=1,2,\ldots,n} \{(1-\lambda\lambda_i)^2\} \|y\|_e^2$$

so that for all $y \in R^n$

$$\|Ay\|_e \le \{\max_{i=1,2,\ldots,n} |1 - \lambda\lambda_i|\} \|y\|.$$

Let i_0 be such that $|1 - \lambda\lambda_{i_0}| \ge |1 - \lambda\lambda_i|$, $i = 1,2,\ldots,n$.
If we let z_{i_0} be a vector with all components zero except
the i_0 component and put $y_{i_0} = Pz_{i_0}$, then

$$\|Ay_{i_0}\|_e = \max_{i=1,2,\ldots,n} \{|1 - \lambda\lambda_i|\} \|y_{i_0}\|.$$

Therefore we conclude that

$$\|A\|_e = \|I - \lambda\frac{\partial^2 f}{\partial x^2}(x^*)\|_e = \max_{i=1,2,\ldots,n} \{|1-\lambda\lambda_i|\}.$$

To have $\|I - \lambda\frac{\partial^2 f}{\partial x^2}(x^*)\|_e < 1$ we clearly need

$$-1 < 1 - \lambda\lambda_i < 1, \qquad i = 1,2,\ldots,n.$$

Since $\frac{\partial^2 f}{\partial x^2}(x^*)$ is positive definite, all λ_i are positive.
Thus the above inequality becomes

$$0 < \lambda < \frac{2}{\lambda_i}, \quad i = 1, 2, \ldots, n.$$

Since λ_1 is the largest of the λ_i, if $0 < \lambda < \frac{2}{\lambda_1}$, all of the required inequalities are satisfied and we have $\left\| \frac{\partial g}{\partial x}(x^*) \right\|_e < 1$. Then by Theorem 7.2 we have

$$\lim_{k \to \infty} x_k = x^*.$$

Now

$$x^* - x_{k+1} = g(x^*) - g(x_k) = \frac{\partial g}{\partial x}(x^*)(x^* - x_k)$$

$$+ \mathcal{O}\|x^* - x_k\|$$

so that

$$\|x^* - x_{k+1}\|_e \leq (\left\| \frac{\partial g}{\partial x}(x^*) \right\|_e + \mathcal{O}(1)) \|x^* - x_k\|$$

$$= (\max_{i=1, 2, \ldots, n} \{(1 - \lambda \lambda_i)\} + \mathcal{O}(1)) \|x^* - x_k\|.$$

If $\max_{i=1, 2, \ldots, n} \{(1 - \lambda \lambda_i)\} < \rho < 1$ then for $\|x^* - x_k\|$ sufficiently small, i.e., for k sufficiently large, we have

$$\|x^* - x_{k+1}\|_e \leq \rho \|x^* - x_k\|_e$$

and the proof is complete.

In most cases it is rather difficult ahead of time to determine the value of λ_1 and one simply tries a reasonably small value of λ. The above theorem shows however that λ must be correspondingly small as $\left\|\frac{\partial^2 f}{\partial x^2}(x^*)\right\|$ is large.

The gradient method will also work if $\frac{\partial^2 f}{\partial x^2}(x^*)$ is only positive semi-definite, provided it is still true that the minimum x^* is an isolated solution of $\frac{\partial f}{\partial x} = 0$. The rate of convergence will normally be quite slow under these circumstances however. No estimate of the form $\|x^* - x_{k+1}\| \leq \rho \|x^* - x_k\|$, $0 < \rho < 1$, will be available.

We now give an example of the use of the gradient method.

EXAMPLE. The Cosmopolitan Encyclopedia Co. normally sells its product for cash or for no money down with cost spread over twenty four equal payments. The basic selling price is the same in either case but a service charge equal to a certain percent of the selling price is added to time payment accounts. Thus the company's income comes from two sources, the price

charged for the encyclopedias and the service charges col-
lected on time payment accounts.

The encyclopedias cost \$100 to produce. We will let
the selling price be \$100$(1 + x)$, where x is, of course,
positive. We let the service charge on time payments be y
percent of the selling price.

Experience indicates the following to be true:
(i) total sales are proportional to $\dfrac{1}{1+x+x^2}$ in the price range
under consideration; (ii) the fraction of total sales which are
cash sales is $\dfrac{y}{20} + \dfrac{20-y}{20}\dfrac{1}{2(1+x)}$ in the range $0 \leq y \leq 20$.
The question is: how should x and y be set so as to realize
a maximum profit?

The total profit is \$$p$ = \$$p_1$ + \$$p_2$, where \$$p_1$ is the
profit realized from the mark-up x and \$$p_2$ is the profit
realized from the service charge y on time payment accounts.
Thus

$$p_1 = \left(\frac{\alpha}{1+x+x^2}\right)100\,x$$

for some positive α. On the other hand

$$p_2 = (\frac{\alpha}{1+x+x^2})(\frac{20-y}{20})(1 - \frac{1}{2(1+x)})100(1+x)y .$$

Thus

$$p = 100\alpha(\frac{x}{1+x+x^2} + \frac{(y-\frac{y^2}{20})(x+\frac{1}{2})}{1+x+x^2})$$

and x and y should be chosen so as to maximize

$$f(x, y) = \frac{x}{1+x+x^2} + \frac{(y-\frac{y^2}{20})(x+\frac{1}{2})}{1+x+x^2} .$$

We compute the partial derivatives of f:

$$\frac{\partial f}{\partial x} = \frac{1}{1+x+x^2}(1+y-\frac{y^2}{20})-\frac{2x+1}{(1+x+x^2)^2}(x+(y-\frac{y^2}{20})(x+\frac{1}{2})),$$

$$\frac{\partial f}{\partial y} = \frac{x+\frac{1}{2}}{1+x+x^2}(1-\frac{y}{10}).$$

In this particular case the equations $\frac{\partial f}{\partial x} = 0$, $\frac{\partial f}{\partial y} = 0$ can be solved by hand. Since $x > 0$, $\frac{\partial f}{\partial y} = 0$ implies $1 - \frac{y}{10} = 0$, i. e., $y = 10$. Then

$$0 = \frac{\partial f}{\partial x} = \frac{6}{1+x+x^2} - \frac{(2x+1)(6x+\frac{5}{2})}{(1+x+x^2)^2}$$

implies $6(1+x+x^2) = (2x+1)(6x+\frac{5}{2})$, i. e. ,

$$6x^2 + 5x - \frac{7}{2} = 0 ,$$

yielding, by the quadratic formula

$$x = \frac{-5 \pm \sqrt{25 + 84}}{12} = .45 \quad \text{or} \quad -1.29 .$$

Clearly the only acceptable value is $x = .45$. Thus the selling price should be \$145 and the service charge should be 10 %.

Let us see now how the gradient method will bring us to the same result. Maximizing $f(x,y)$ is the same as minimizing $-f(x,y)$. The gradient method for minimizing $-f(x,y)$ is

$$x_{k+1} = x_k - \lambda \frac{\partial(-f)}{\partial x}(x_k,y_k) = x_k + \lambda \frac{\partial f}{\partial x}(x_k,y_k) ,$$

$$y_{k+1} = y_k - \lambda \frac{\partial(-f)}{\partial y}(x_k,y_k) = y_k + \lambda \frac{\partial f}{\partial x}(x_k,y_k) .$$

Let us start with $x_0 = 1$, $y_0 = 5$ and take $\lambda = \frac{1}{2}$. Thus

$$x_0 = 1, \quad y_0 = 5$$

$$x_{k+1} = x_k + \frac{1}{2}\left[\frac{1}{1+x_k+x_k^2}\left(1+y_k-\frac{y_k^2}{20}\right)\right.$$

$$\left.-\frac{2x_k+1}{(1+x_k+x_k^2)^2}\left(x_k+\left(y_k-\frac{y_k^2}{20}\right)\left(x_k+\frac{1}{2}\right)\right)\right]$$

$$y_{k+1} = y_k + \frac{1}{2}\left[\frac{x_k+\frac{1}{2}}{1+x_k+x_k^2}\left(1-\frac{y_k}{10}\right)\right], \qquad k \geq 0.$$

We compute

$$x_1 = 1 + \frac{1}{2}\left[\frac{1}{1+1+1^2}\left(1+5-\frac{5^2}{20}\right)-\frac{2\cdot1+1}{(1+1+1^2)^2}\left(1+\left(5-\frac{5^2}{20}\right)\left(1+\frac{1}{2}\right)\right)\right]$$

$$= 1 - \frac{5}{16} = .68750$$

$$y_1 = 5 + \frac{1}{2}\left[\frac{1+\frac{1}{2}}{1+1+1^2}\left(1-\frac{5}{10}\right)\right]$$

$$= 5 + \frac{1}{8} = 5.12500 .$$

We continue in this manner and obtain the following table of values for x_k, y_k:

Table 1 $(\lambda = \frac{1}{2})$

k	x_k	y_k	k	x_k	y_k
0	1.00000	5.00000	88	.45348	9.61328
1	.68750	5.12500	89	.45347	9.62439
2	.47438	5.25900	90	.45346	9.63518
3	.47473	5.39491	91	.45346	9.64567
4	.47236	5.52693	92	.45345	9.65585
5	.47149	5.65519	93	.45345	9.66574
6	.47003	5.77979	94	.45344	9.67534
7	.46904	5.90084	95	.45344	9.68467
8	.46793	6.01843	96	.45343	9.69373
9	.46701	6.13266	97	.45343	9.70253
10	.46609	6.24362	98	.45342	9.71108
11	.46529	6.35142	99	.45342	9.71938
12	.46451	6.45612	100	.45342	9.72745
⋮	⋮	⋮	⋮	⋮	⋮

After much labor the computer is getting us into the
vicinity of the maximum that we seek. It is clear, however,
that the convergence is discouragingly slow. It might seem
reasonable to try to speed things up by increasing λ from
$\frac{1}{2}$ to 1. This change simply leads to disaster as the next
table indicates.

Table 2 (λ = 1)

k	x_k	y_k	k	x_k	y_k
0	1.00000	5.00000	40	.93697	9.34452
1	.37500	5.25000	41	1.29807	9.36840
2	.72277	5.52423	42	.47544	9.39691
3	.22645	5.76799	43	.39606	9.43149
4	1.36639	6.00860	44	.63630	9.46429
5	.67691	6.18457	45	.13155	9.49411
6	.20057	6.39488	46	2.20292	9.52192
7	1.55635	6.59843	47	1.63960	9.53796
8	.87845	6.73893	48	.90732	9.55652
9	.17923	6.90855	49	.10743	9.57938
10	1.72837	7.08189	50	2.41572	9.60221
⋮	⋮	⋮	⋮	⋮	⋮

We have succeeded in speeding up the convergence of the y_k toward 10 but the x_k values now show no sign of convergence at all!

This perplexing behavior is very common and represents a serious limitation on the usefulness of the gradient method. Let us analyze this difficulty and then see what we can do to overcome it.

In order to study the difficulties inherent in the

gradient method let us consider the minimization of $f: R^n \to$ R^1 given by

$$f(x) = x'Ax$$

where A is an $n \times n$ symmetric positive definite matrix. The minimum obviously occurs at $x = 0$. The gradient method is

$$x_{k+1} = x_k - \lambda \frac{\partial f}{\partial x}(x_k)' = x_k - 2\lambda Ax_k$$

$$= (I - 2\lambda A)x_k.$$

From Chapter 7 we know that there exists an orthogonal matrix P which diagonalizes A, i. e.,

$$P^{-1}AP = P'AP = \text{diag}(\lambda_1, \lambda_2, \ldots, \lambda_n).$$

Setting $x_k = Py_k$, $k = 0, 1, 2, \ldots$ we have

$$Py_{k+1} = (I - 2\lambda A)Py_k$$

or multiplying by P^{-1}

$$y_{k+1} = (I - 2\lambda P^{-1}AP)y_k$$

$$= \text{diag}(1 - 2\lambda\lambda_1, 1 - 2\lambda\lambda_2, \ldots, 1 - 2\lambda\lambda_n)y_k.$$

For the components of the vectors y_k we thus have

$$y_{k+1}^j = (1 - 2\lambda\lambda_j)y_k^j, \quad j = 1, 2, \ldots, n.$$

Thus

$$y_k^j = (1 - 2\lambda\lambda_j)^k y_0^j, \quad j = 1, 2, \ldots, n, \quad k = 0, 1, 2, \ldots$$

We see that the y_k^j form a geometric progression y_0^j,

$(1 - 2\lambda\lambda_j)y_0^j$, $(1 - 2\lambda\lambda_j)^2 y_0^j, \ldots$.

The rate at which these numbers approach zero is

clearly dependent upon $|1 - 2\lambda\lambda_j|$. We see from Theorem 8.3

that if λ_1 is the largest eigenvalue of $A = \dfrac{1}{2}\dfrac{\partial^2 f}{\partial x^2}$ we obtain

convergence if $0 < \lambda < \dfrac{1}{\lambda_1}$. But the <u>rate</u> of convergence

depends upon the quantity $\max\limits_{j=1,2,\ldots,n} \{|1 - 2\lambda\lambda_j|\}$. Suppose

λ_n is the smallest eigenvalue of A and λ satisfies $0 <$

$\lambda < \dfrac{1}{\lambda_1}$. Then if $\lambda_n \leq \dfrac{\lambda_1}{2}$

$$|1 - 2\lambda\lambda_n| = 1 - 2\lambda\lambda_n \geq 1 - 2\frac{\lambda_n}{\lambda_1}.$$

If λ_n is much smaller than λ_1 this quantity is very close

to 1 and y_k^n will converge to zero very slowly.

The analysis of the convergence of the gradient method

to x^*, the minimum of an arbitrary twice continuously dif-

ferentiable function f, is much the same. If the ratio of the

smallest eigenvalue of $\dfrac{\partial^2 f}{\partial x^2} (x^*)$ to the largest eigenvalue of

that matrix is very small we can expect slow convergence.

There are several ways to get around this problem.

One possibility is to use the gradient method to obtain a

rough approximation to x^* and then use Newton's method to

calculate high precision approximations to x^* if these are

desired. As indicated in Exercise 3 at the end of this

chapter, Newton's method converges much more rapidly than

the gradient method, provided one starts near x^*. The

gradient method works better than Newton's method at the job

of locating the vicinity of a local minimum from distant start-

ing points, as we will explain shortly.

The problem with Newton's method, $x_{k+1} = x_k -$

$\dfrac{\partial^2 f}{\partial x^2}(x_k)^{-1}\dfrac{\partial f}{\partial x}(x_k)'$ is that it requires computation of the n^2

entries of the Hessian matrix $\dfrac{\partial^2 f}{\partial x^2}(x_k)$ and then one must

solve the equation $\dfrac{\partial^2 f}{\partial x^2}(x_k)(x_{k+1} - x_k) = -\dfrac{\partial f}{\partial x}(x_k)'$. If the

expressions for these second partial derivatives are quite

complicated it may require considerable computation time

just to calculate these partial derivatives and it may well

turn out that little, if any, time is saved by the use of

Newton's method.

Let us consider now a modification of the gradient

method. Instead of subtracting a scalar multiple of $\frac{\partial f}{\partial x}(x_k)'$

from x_k we use a matrix, i. e. , we set

$$x_{k+1} = x_k - P\frac{\partial f}{\partial x}(x_k)' \ .$$

If we define $g(x) = x - P\frac{\partial f}{\partial x}(x)'$ then

$$\frac{\partial g}{\partial x} = I - P\frac{\partial^2 f}{\partial x^2} \ .$$

If P can be chosen close to $\frac{\partial^2 f}{\partial x^2}(x^*)^{-1}$ then $\|\frac{\partial g}{\partial x}\|$ will be

very small near x^* and, following the proof of Theorem 7. 2,

we will obtain rapid convergence. One way to obtain P is

to evaluate $\frac{\partial^2 f}{\partial x^2}(\hat{x})$ for some point $\hat{x}$ fairly close to x^*

and take $P = \frac{\partial^2 f}{\partial x^2}(\hat{x})^{-1}$ Clearly what we are doing here is

trying to come close to Newton's method without having to

evaluate a completely new matrix $\frac{\partial^2 f}{\partial x^2}(x_k)$ at each step. We

still have the problem of calculating and evaluating the second

order partial derivatives of f. This can be a tedious process as the reader can verify for himself in even so simple a situation as the encyclopedia pricing problem above.

Fortunately, it turns out that we can use this idea without actually evaluating $\frac{\partial^2 f}{\partial x^2}$. Since f is assumed twice continuously differentiable we have, for x near x^*

$$\frac{\partial f}{\partial x}(x)' = \frac{\partial f}{\partial x}(x^*)' + \frac{\partial}{\partial x}\left(\frac{\partial f}{\partial x}\right)(x^*)(x - x^*)$$

$$+ \mathcal{O}(\|x - x^*\|) = \frac{\partial^2 f}{\partial x^2}(x^*)(x - x^*) + \mathcal{O}(\|x - x^*\|)$$

since $\frac{\partial f}{\partial x}(x^*) = 0$. Substituting this into the gradient method formula $x_{k+1} = x_k - \lambda \frac{\partial f}{\partial x}(x_k)'$ we have

$$x_{k+1} = x_k - \lambda \frac{\partial^2 f}{\partial x^2}(x^*)(x_k - x^*) + \mathcal{O}(\|x_k - x^*\|).$$

Repeating, with k replaced by k+1, we obtain

$$x_{k+2} = x_{k+1} - \lambda \frac{\partial^2 f}{\partial x^2}(x^*)(x_{k+1} - x_k) + \mathcal{O}(\|x_{k+1} - x^*\|).$$

Ignoring the "$\mathcal{O}$" terms

$$x_{k+2} - x_{k+1} \approx [I - \lambda \frac{\partial^2 f}{\partial x^2}(x^*)](x_{k+1} - x_k),$$

where "$\approx$" means "approximately equal to". Thus

$$\frac{\partial^2 f}{\partial x^2}(x^*)(x_{k+1} - x_k) \approx \frac{1}{\lambda}[x_{k+1} - x_k - x_{k+2} + x_{k+1}]$$

and if we define

$$\Delta x_k = x_{k+1} - x_k ,$$

$$\Delta^2 x_k = x_{k+2} - 2x_{k+1} + x_k ,$$

we have

$$\frac{\partial^2 f}{\partial x^2}(x^*)\Delta x_k \approx -\frac{1}{\lambda}\Delta^2 x_k$$

or, equivalently,

$$\frac{\partial^2 f}{\partial x^2}(x^*)^{-1}\Delta^2 x_k = -\lambda \Delta x_k .$$

Suppose the vectors $\Delta^2 x_1, \Delta^2 x_2, \dots, \Delta^2 x_n$ are linearly independent. Then for any n-vector y we have unique real numbers $\alpha^1, \alpha^2, \dots, \alpha^n$ such that

$$y = \alpha^1 \Delta^2 x_1 + \alpha^2 \Delta^2 x_2 + \dots + \alpha^n \Delta^2 x_n$$

and

$$\frac{\partial^2 f}{\partial x^2}(x^*)^{-1}y = \frac{\partial^2 f}{\partial x^2}(x^*)^{-1}[\alpha^1\Delta^2 x_1 + \alpha^2\Delta^2 x_2 + \ldots + \alpha^n\Delta^2 x_n]$$

$$\approx -\lambda[\alpha^1\Delta x_1 + \alpha^2\Delta x_2 + \ldots + \alpha^n\Delta x_n] .$$

Thus, using $\Delta x_1, \Delta x_2, \ldots, \Delta x_n, \Delta^2 x_1, \Delta^2 x_2, \ldots, \Delta^2 x_n$ we

can compute $\dfrac{\partial^2 f}{\partial x^2}(x^*)^{-1}y$ approximately for any $y \in R^n$.

Another way of putting this is to say that the linear trans-

formation P which has the property that

$$P\Delta^2 x_k = -\lambda\Delta x_k$$

is a good approximation to $\dfrac{\partial^2 f}{\partial x^2}(x^*)$. We can then use the

method

$$\hat{x}_{k+1} = \hat{x}_k - P\frac{\partial f}{\partial x}(\hat{x}_k)'$$

and expect that the resulting sequence $\{\hat{x}_k\}$ will converge

quite rapidly to x^*.

To compute P we need $\Delta x_k, \Delta^2 x_k, k = 1, 2, \ldots, n$.

These require for their computation the vectors $x_1, x_2, \ldots, x_n$,

x_{n+1}, x_{n+2} obtained by the use of the ordinary gradient method

$x_{k+1} = x_k - \lambda\dfrac{\partial f}{\partial x}(x_k)'$. Now in actual fact one could just as

well use later vectors in the sequence, $x_{j+1}, x_{j+2}, \ldots, x_{j+n}$,

$x_{j+n+1}, x_{j+n+2},$ to compute $\Delta x_{j+k}, \Delta^2 x_{j+k},$ $k = 1, 2, \ldots, n$.

In any case, letting $(\Delta x_{j+1}, \Delta x_{j+2}, \ldots, \Delta x_{j+n})$ denote the

$n \times n$ matrix whose kth column is Δx_{j+k} and treating $\Delta^2 x_{j+k}$

similarly we have

$$P(\Delta^2 x_{j+1}, \Delta^2 x_{j+2}, \ldots, \Delta^2 x_{j+n}) = -\lambda(\Delta x_{j+1}, \Delta x_{j+2}, \ldots, \Delta x_{j+n})$$

and, assuming $\Delta x_{j+1}, \Delta x_{j+2}, \ldots, \Delta x_{j+n}$ are linearly inde-

pendent

$$P = -\lambda (\Delta x_{j+1}, \Delta x_{j+2}, \ldots, \Delta x_{j+n})(\Delta^2 x_{j+1}, \Delta^2 x_{j+2}, \ldots, \Delta^2 x_{j+n})^{-1}.$$

In the case of the encyclopedia problem we have $n = 2$.

Taking $j = 5$ we construct the following table from Table 1:

Table 3

k	x_k	Δx_k	$\Delta^2 x_k$	y_k	Δy_k	$\Delta^2 y_k$
5	.47149	-.00149	.00050	5.65519	.12460	-.00355
6	.47003	-.00099	-.00012	5.77979	.12105	-.00346
7	.46904	-.00111		5.90084	.11759	
8	.46793			6.01843		

Thus

$$P = -\frac{1}{2} \begin{pmatrix} -.00149 & -.00099 \\ .12460 & .12105 \end{pmatrix} \begin{pmatrix} .00050 & -.00012 \\ -.00355 & -.00346 \end{pmatrix}^{-1}$$

$$-\frac{1}{2} \begin{pmatrix} -.00149 & -.00099 \\ .12460 & .12105 \end{pmatrix} \begin{pmatrix} 1600 & -49 \\ -1600 & -230 \end{pmatrix} = \begin{pmatrix} .40 & -.15 \\ -2.8 & 17 \end{pmatrix}.$$

Note that because of the loss of significant figures due to subtraction of nearly equal numbers to obtain the Δx_k, Δy_k, $\Delta^2 x_k$, $\Delta^2 y_k$ we are not justified in specifying P more precisely than we have. Since P is only an approximation to $\frac{\partial^2 f}{\partial x^2}(x^*)$ there is no point in trying for greater precision.

The modified gradient method for the encyclopedia pricing problem is therefore

$$\begin{pmatrix} \hat{x}_{k+1} \\ \hat{y}_{k+1} \end{pmatrix} = \begin{pmatrix} \hat{x}_k \\ \hat{y}_k \end{pmatrix} + \begin{pmatrix} .40 & -.15 \\ -2.8 & 17 \end{pmatrix} \begin{pmatrix} \frac{\partial f}{\partial x}(\hat{x}_k, \hat{y}_k) \\ \frac{\partial f}{\partial y}(\hat{x}_k, \hat{y}_k) \end{pmatrix}.$$

Let us begin where Table 3 leaves off, i. e. , we put

$$x_8 = x_8 = .46793,$$

$$y_8 = y_8 = 6.01843.$$

Substituting these values into the expressions for $\frac{\partial f}{\partial x}$, $\frac{\partial f}{\partial y}$ we obtain

$$\begin{pmatrix} \hat{x}_9 \\ \hat{y}_9 \end{pmatrix} = \begin{pmatrix} .46793 \\ 6.01843 \end{pmatrix} + \begin{pmatrix} .40 & -.15 \\ -2.8 & 17 \end{pmatrix} \begin{pmatrix} -.00184 \\ .22846 \end{pmatrix}$$

$$= \begin{pmatrix} .46793 \\ 6.01843 \end{pmatrix} + \begin{pmatrix} -.03501 \\ 3.88897 \end{pmatrix} = \begin{pmatrix} .43292 \\ 9.90740 \end{pmatrix}.$$

As far as the y component is concerned, $\hat{y}_9$ is a better approximation to $y^* = 10$ than is y_{100}! If we were to continue with this method we would obtain a sequence $\{(\begin{smallmatrix} \hat{x}_k \\ \hat{y}_k \end{smallmatrix})\}$ converging very rapidly to the maximum $(\begin{smallmatrix} x^* \\ y^* \end{smallmatrix})$.

There is one point we should make before leaving this subject. Since the Hessian of f is a symmetric matrix, so is its inverse. As we have seen in the above example, our approximation P may not be symmetric. The entries p_2^1 and p_1^2 are both approximations to $\dfrac{\partial^2 f}{\partial x \partial y}$ but are arrived at in different ways and thus may not coincide. It makes sense to average these two. Thus we could just as well use

$$P = \begin{pmatrix} .40 & -1.46 \\ -1.46 & 17 \end{pmatrix}$$

rather than the matrix used above. In general we may as well replace P by $\dfrac{1}{2}(P + P')$ when approximating the symmetric matrix $\dfrac{\partial^2 f}{\partial x^2} (x^*)^{-1}$.

The process which we have used above to obtain an

approximation P to $\dfrac{\partial^2 f}{\partial x^2} (x^*)^{-1}$ is essentially what is com-

monly called <u>Aitken's Δ^2-process</u>. It should only be used

after the gradient method has brought the vectors x_k within

a fairly small neighborhood of x^*. By "fairly small neighbor-

hood", we mean a neighborhood in which the Hessian of f

is not very different from $\dfrac{\partial^2 f}{\partial x^2} (x^*)$. The Δ^2-process should

be viewed, just as Newton's process is viewed, as a rapid

"polishing off" procedure to be used after the standard

gradient method has approximately located x^*.

It is time now to give some consideration to the ques-

tion of <u>global</u> convergence of the gradient method. Up to the

present we have only asked whether or not $\lim\limits_{k \to \infty} x_k = x^*$

when x_0 is chosen sufficiently close to x^*. When we con-

sider global convergence the appropriate problem is that of

characterizing as large as possible a set of initial points x_0

for which $\lim\limits_{k \to \infty} x_k = x^*$ when λ is sufficiently small. In

order to do this we need to develop some further background

material.

DEFINITION 8.1. Let $f : R^n \rightarrow R^1$ be a continuous function.
We define subsets $S_c \subseteq R^n$ for each real number c by

$$S_c = \{x \in R^n | f(x) = c\}.$$

The set S_c is called the <u>c-contour surface</u> of f.

EXAMPLES. Let $n = 2$ and consider $f : R^2 \rightarrow R^1$ given by

$$f(x, y) = x^2 + xy + y^2 .$$

The set $S_c = \{ \binom{x}{y} \in R^2 | x^2 + xy + y^2 = c \}$ is empty for
$c < 0$, the single point $\binom{0}{0}$ for $c = 0$ and an ellipse for
$c > 0$. These contour surfaces are shown in Figure 13.

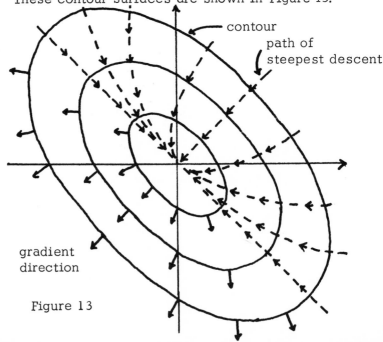

contour

path of
steepest descent

gradient
direction

Figure 13

Note how these contours form a set of nested closed curves
surrounding the minimum $\begin{pmatrix} x^* \\ y^* \end{pmatrix} = \begin{pmatrix} 0 \\ 0 \end{pmatrix}$.

Again let $n = 2$ and define $f : R^2 \rightarrow R^1$ by

$$f(x,y) = (x+1)^2(x-1)^2 + y^2 = (x^2-1)^2 + y^2.$$

Again S_c is empty for $c < 0$. For $c = 0$ S_c consists of
two points, $\begin{pmatrix} 1 \\ 0 \end{pmatrix}$ and $\begin{pmatrix} -1 \\ 0 \end{pmatrix}$. For $0 < c < 1$ S_c consists of
two closed curves surrounding the points $\begin{pmatrix} 1 \\ 0 \end{pmatrix}$ and $\begin{pmatrix} 0 \\ 1 \end{pmatrix}$.
For $c = 1$, S_c has the form of a figure-eight. For $c > 1$
S_c has a sort of dumb-bell shape. All these are shown in
Figure 14.

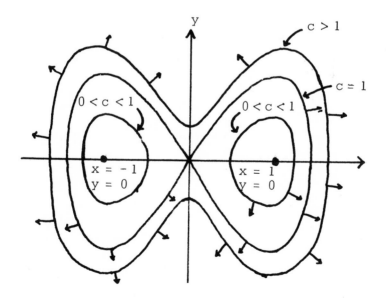

Figure 14

In higher dimensions we have analogous behavior. In R^3 contour surfaces are two dimensional surfaces. If $x^* \in R^3$ is an isolated local minimum or local maximum for f then near x^* the surfaces S_c form closed "balloons" with x^* in the interior. Away from x^* their shapes may become very complicated and a complete description of their behavior may be out of the question.

In Chapter 9 we will show that if $f : R^n \to R^1$ is continuously differentiable and if $\hat{x} \in R^n$ is a point where $\frac{\partial f}{\partial x}(\hat{x}) \neq 0$ then the vector $\frac{\partial f}{\partial x}(\hat{x})'$ lies perpendicular to $S_{f(\hat{x})}$ at $\hat{x}$ and points in the direction of maximum increase of f. The arrows in Figures 13 and 14 indicate the direction of the transposed gradient vector, $\begin{pmatrix} \frac{\partial f}{\partial x} \\ \frac{\partial f}{\partial y} \end{pmatrix}$. Note that these vectors point from S_c out toward surfaces $S_{\hat{c}}$ with $\hat{c} > c$.

Consider now a vector differential equation

$$\frac{dx}{dt} = g(x)$$

where $g : R^n \to R^n$ is continuous. A solution of this differential equation corresponding to the initial point x_0 is a differentiable function $x : R^1 \to R^n$ such that

$$\frac{dx^i(t)}{dt} \equiv g^i(x(t)), \qquad i = 1, 2, \ldots, n, \quad 0 \leq t < T,$$

$$x(0) = x_0.$$

With g continuous it can be shown that such an <u>initial value problem</u> has at least one solution $x(t)$ defined on some interval $[0, T)$, $T > 0$. If g is continuously differentiable there is exactly one solution.

As t varies in $[0, T)$, $x(t)$ traces out a curve $\tilde{x}$ in R^n. Because $x(t)$ satisfies the differential equation shown above we can verify that at each point $x(t)$ lying on the curve $\tilde{x}$, the vector $g(x(t))$ is <u>tangent</u> to the curve $\tilde{x}$, provided $g(x(t)) \neq 0$. (We will define the word <u>tangent</u> more precisely in Chapter 9.) As t increases the point $x(t)$ moves along $\tilde{x}$ in the direction indicated by the vector $g(x(t))$.

Suppose $f : R^n \to R^1$ is continuously differentiable. Then we can put $g = -\dfrac{\partial f'}{\partial x}$ and consider the differential equation

$$\frac{dx}{dt} = -\frac{\partial f}{\partial x}(x)'.$$

Solutions $x(t)$ of this differential equation give rise to

curves $\tilde{x}$ in R^n with the property that if $x(t) \in \tilde{x}$, then

$-\dfrac{\partial f}{\partial x}(x(t))$ is tangent to the curve $\tilde{x}$ at the point $x(t)$. For

each fixed t, $-\dfrac{\partial f}{\partial x}(x(t))'$ lies perpendicular to the contour

surface $S_{f(x(t))}$ at $x(t)$ and points in the direction of most

rapid decrease of f. Therefore the curve $\tilde{x}$ lies perpendi-

cular to each of the surfaces S_c which it crosses and as t

increases the point $x(t)$ moves along $\tilde{x}$ in the direction

$-\dfrac{\partial f}{\partial x}(x(t))'$, the direction of greatest decrease of f. For this

reason $\tilde{x}$ is called a <u>path of steepest descent</u> for the func-

tion f.

EXAMPLE. In R^2 we put $f(x,y) = x^2 + xy + y^2$. The ap-

propriate vector differential equation is

$$\frac{d}{dt}\begin{pmatrix} x \\ y \end{pmatrix} = \begin{pmatrix} -\dfrac{\partial f}{\partial x}(x,y) \\ -\dfrac{\partial f}{\partial y}(x,y) \end{pmatrix} = \begin{pmatrix} -2x - y \\ -2y - x \end{pmatrix},$$

i. e.

$$\frac{d}{dt}\begin{pmatrix} x \\ y \end{pmatrix} = \begin{pmatrix} -2 & -1 \\ -1 & -2 \end{pmatrix}\begin{pmatrix} x \\ y \end{pmatrix}.$$

To solve this vector differential equation we find it conven-

ient to diagonalize the matrix $\begin{pmatrix} -2 & -1 \\ -1 & -2 \end{pmatrix}$. Putting

$$\begin{pmatrix} x \\ y \end{pmatrix} = \begin{pmatrix} 1 & 1 \\ -1 & 1 \end{pmatrix} \begin{pmatrix} w \\ z \end{pmatrix} \quad \text{or} \quad \begin{pmatrix} w \\ z \end{pmatrix} = \frac{1}{2} \begin{pmatrix} 1 & -1 \\ 1 & 1 \end{pmatrix} \begin{pmatrix} x \\ y \end{pmatrix}$$

we have

$$\frac{d}{dt} \left[\begin{pmatrix} 1 & 1 \\ -1 & 1 \end{pmatrix} \begin{pmatrix} w \\ z \end{pmatrix} \right] = \begin{pmatrix} -2 & -1 \\ -1 & -2 \end{pmatrix} \begin{pmatrix} 1 & 1 \\ -1 & 1 \end{pmatrix} \begin{pmatrix} w \\ z \end{pmatrix}$$

or

$$\frac{d}{dt} \begin{pmatrix} w \\ z \end{pmatrix} = \frac{1}{2} \begin{pmatrix} 1 & -1 \\ 1 & 1 \end{pmatrix} \begin{pmatrix} -2 & -1 \\ -1 & -2 \end{pmatrix} \begin{pmatrix} 1 & 1 \\ -1 & 1 \end{pmatrix} \begin{pmatrix} w \\ z \end{pmatrix} =$$

$$= \begin{pmatrix} -1 & 0 \\ 0 & -3 \end{pmatrix} \begin{pmatrix} w \\ z \end{pmatrix}$$

which gives us

$$\frac{dw}{dt} = -w, \qquad \frac{dz}{dt} = -3z.$$

The general solutions of these equations are of the form

$$w(t) = w_0 e^{-t}, \qquad z(t) = z_0 e^{-3t}.$$

Since $e^{-3t} = (e^{-t})^3$ it follows that

$$z(t) = c(w(t))^3$$

for some real constant c, unless $w_0 = 0$, in which case $w(t) \equiv 0$.

Thus in the original variables x, y the paths of steepest descent for the function $f(x, y) = x^2 + xy + y^2$ are described by equations

$$\frac{x+y}{2} = c\left(\frac{x-y}{2}\right)^3$$

or, since c is an arbitrary real constant

$$x + y = \hat{c}(x - y)^3.$$

The solution $w(t) \equiv 0$ corresponds to the curve $x = y$, also a path of steepest descent. The contour curves and paths of steepest descent for $x^2 + xy + y^2$ are shown in Figure 13. The arrows indicate the descent directions.

An explorer seeking to find the lowest point in a valley might well try to find this point simply by heading "downhill" until this is no longer possible. The paths of steepest descent represent "downhill" directions as far as the graph of f is concerned. The gradient method can be viewed as an attempt to follow these "downhill" directions in a sense which we are going to make precise.

Now let us return to differential equations $\frac{dx}{dt} = g(x)$,

$g : R^n \to R^n$. Assuming that g is continuously differentiable

there will be exactly one solution $x(t)$, defined on an inter-

val $[0, T)$, $T > 0$, corresponding to a given initial point x_0.

A common technique for approximating $x(t)$ on a closed in-

terval $[0, T) \subset [0, T)$ is known as <u>Euler's method</u>.

Let λ be a positive real number and set $t_k = k\lambda$,

$k = 0, 1, 2, \ldots$. Since $x(t)$ satisfies $\frac{dx}{dt} = g(x)$ we have

$$x(t_{k+1}) - x(t_k) = \int_{t_k}^{t_{k+1}} \frac{dx}{dt} dt = \int_{t_k}^{t_{k+1}} g(x(t)) dt .$$

If the interval $[t_k, t_{k+1}]$ is small, i.e., if λ is small,

the constant $g(x(t_k))$ is a fairly good approximation to

$g(x(t))$ in this interval. Then

$$x(t_{k+1}) - x(t_k) \approx \int_{t_k}^{t_{k+1}} g(x(t_k)) dt = \lambda g(x(t_k))$$

i. e.

$$x(t_{k+1}) \approx x(t_k) + \lambda g(x(t_k)), \quad \lambda = t_{k+1} - t_k .$$

Therefore it makes sense to try to approximate the vectors

$x(t_k)$, $k = 0, 1, 2, \ldots$ by vectors x_k, where x_0 is the given

initial point and

$$x_{k+1} = x_k + \lambda g(x_k), \qquad k = 0, 1, 2, \ldots .$$

This is Euler's method for approximating $x(t_k)$, $k = 0, 1, 2, \ldots$.

We state the following theorem without proof.   The proof may

be found in any good numerical analysis text.

THEOREM 8. 4.   Let $g : R^n \to R^n$ be continuously differenti-

able and let $x(t)$ satisfy $\frac{dx}{dt} = g(x)$ on the interval $[0, T_1]$,

with $x(0) = x_0$.   Let the sequence of vectors $\{x_k\}$ be gen-

erated by Euler's method.   Then there are positive numbers

λ_0 and K such that whenever $0 < \lambda \leq \lambda_0$ we have, for

$t_k = k\lambda$,

$$\max \{\|x(t_k) - x_k\| \,|\, t_k \in [0, T_1]\} \leq K\lambda .$$

Thus Euler's method is <u>convergent</u> in the sense that

$$\lim_{\lambda \to 0} [\max \{\|x(t_k - x_k\| \,|\, t_k \in [0, T_1]\}] = 0 .$$

Using Theorem 8. 4 we can obtain two results which

provide a partial solution to the global convergence problem

for the gradient method.

THEOREM 8. 5.   Let $f : R^n \rightarrow R^1$ be twice continuously dif-

ferentiable with a local minimum at $x^* \epsilon R^n$ where $\dfrac{\partial^2 f}{\partial x^2}(x^*)$

is positive definite.   Let $x_0 \epsilon R^n$ and assume there is a

path of steepest descent from x_0 to x^*, i. e. , there is a

solution $x(t)$ of $\dfrac{dx}{dt} = -\dfrac{\partial f'}{\partial x}$ defined on $[\,0\,,\infty)$ with

$\lim\limits_{t \to \infty} x(t) = x^*$.   Then the gradient method, starting at x_0,

converges to x^* if λ is sufficiently small.

Proof.   From Theorem 8. 3 there is an $\epsilon > 0$ such that the

gradient method converges to x^*, provided we start within

the neighborhood $N(x^*, \epsilon) = \{x \epsilon R^n \mid \|x^* - x\| < \epsilon \}$.   Con-

sequently all we have to do is to show that if λ is suffi-

ciently small there is a positive integer k_0 such that x_{k_0}

lies in $N(x^*, \epsilon)$.   Since $\lim\limits_{t \to \infty} x(t) = x^*$ there is a positive

number T_1 such that $\|x^* - x(t)\| < \epsilon/2$ for $t > T_1 - 1$.

Now the gradient method is just Euler's method ap-

plied to the differential equation $\dfrac{dx}{dt} = -\dfrac{\partial f'}{\partial x}$.   According to

Theorem 8. 4, then,

$$\max \{ \|x(t_k) - x_k\| \, | \, t_k \in [0, T_1] \} \leq K\lambda ,$$

provided $0 < \lambda \leq \lambda_0$. Let

$$\lambda_1 = \min \{1, \lambda_0, \frac{\epsilon}{2K} \} .$$

Let $0 < \lambda \leq \lambda_1$ and let k_0 be the largest integer such that $t_{k_0} = k_0 \lambda$ lies in $[0, T_1]$. Then certainly $T_1 - t_{k_0} < \lambda \leq 1$ so $t_{k_0} > T_1 - 1$ and we have $\|x^* - x(t_{k_0})\| < \epsilon /2$. Since $0 < \lambda \leq \lambda_1 \leq \lambda_0$ we have

$$\|x(t_{k_0}) - x_{k_0}\| \leq K\lambda \leq K\frac{\epsilon}{2K} = \frac{\epsilon}{2} .$$

Then

$$\|x^* - x_{k_0}\| \leq \|x^* - x(t_{k_0})\| + \|x(t_{k_0}) - x_{k_0}\| < \frac{\epsilon}{2} + \frac{\epsilon}{2} = \epsilon .$$

Theorem 8. 3 then guarantees that the gradient method will proceed from x_{k_0} and converge to x^* and the proof is complete.

Again we remark that the condition that $\frac{\partial^2 f}{\partial x^2}(x^*)$ be positive definite is not really necessary. We have included it here because we proved Theorem 8.3 with this assumption.

As we noted following Theorem 8. 3, local convergence is still obtained if $\frac{\partial^2 f}{\partial x^2}(x^*)$ is positive semi-definite (which it must be if x^* is to be a minimum) and x^* is an isolated solution of $\frac{\partial f}{\partial x} = 0$. Theorem 8. 5 will also hold under these circumstances.

THEOREM 8. 6. Let $f : R^n \to R^1$ be twice continuously differentiable and assume

(i) x^* is a global minimum for f in R^n;

(ii) $\frac{\partial f}{\partial x}(\hat{x}) \neq 0$ if $\hat{x} \neq x^*$;

(iii) $\lim\limits_{\|x\| \to \infty} f(x) = \infty$.

Then, given any $x_0 \in R^n$, if λ is sufficiently small, the gradient method starting at x_0 yields a sequence of vectors $\{x_k\}$ with

$$\lim_{k \to \infty} x_k = x^* .$$

Proof. From Theorem 8. 5 and the remark following it, all we need to do is to show that if $x(t)$ is the solution of $\frac{dx}{dt} = -\frac{\partial f'}{\partial x}$

with $x(0) = x_0$, then $x(t)$ is defined for $0 \leq t < \infty$ and satisfies

$$\lim_{t \to \infty} x(t) = x^*.$$

The proof that the solution $x(t)$ remains defined for all positive t lies outside the scope of this text. Such a proof may be found in standard differential equations texts. Here we will assume that result and content ourselves with showing that $\lim_{t \to \infty} x(t) = x^*$.

First of all, $f(x(t))$ is monotone decreasing, for if $\hat{t} > \tilde{t}$

$$f(x(\hat{t})) - f(x(\tilde{t})) = \int_{\tilde{t}}^{\hat{t}} \frac{d}{dt} f(x(t)) dt = \int_{\tilde{t}}^{\hat{t}} \frac{\partial f}{\partial x}(x(t)) \frac{dx}{dt} dt$$

$$= \int_{\tilde{t}}^{\hat{t}} \frac{\partial f}{\partial x}(x(t))(-\frac{\partial f}{\partial x}(x(t))') dt = -\int_{\tilde{t}}^{\hat{t}} \left\| \frac{\partial f}{\partial x}(x(t))' \right\|_e^2 dt \leq 0.$$

If we let $S = \{x \in R^n \mid f(x) \leq f(x_0)\}$ then it is clear that $x(t) \in S$ for all $t \geq 0$. From (iii) we see that S is bounded. Since f is continuous, S is closed.

Suppose it were not true that $\lim_{t \to \infty} x(t) = x^*$. Then there would be an $\epsilon > 0$ such that $\|x^* - x(t)\| \geq \epsilon$ for all

$t \geq 0$. Then for all $t \geq 0$ $x(t) \in S_\epsilon$ where S_ϵ is the compact set

$$S_\epsilon = S \cap \{x \in R^n \mid \|x^* - x\| \geq \epsilon\}.$$

The continuous function $h(x) = \|\frac{\partial f}{\partial x}(x)'\|_e^2$ is, by (ii), never

zero in S_ϵ. Since S_ϵ is compact there must be a $\delta > 0$

such that

$$\left\|\frac{\partial f}{\partial x}(x)'\right\|_e^2 \geq \delta, \quad x \in S_\epsilon.$$

But then for all $t \geq 0$

$$f(x(t)) - f(x(0)) = -\int_0^t \left\|\frac{\partial f}{\partial x}(x(t))'\right\|_e^2 dt \leq -\delta t$$

and we conclude that $\lim_{t \to \infty} f(x(t)) = -\infty$. This contradicts

our definition of x^* and must be false. We conclude, there-

fore, that $\lim_{t \to \infty} x(t) = x^*$ and the proof is complete.

EXERCISES. CHAPTER 8

1.   Let $f : R^2 \to R^1$ be defined by

$$f(x, y) = (x - x^*)^4 + (y - y^*)^4,$$

so that f has a unique global minimum at $\begin{pmatrix} x^* \\ y^* \end{pmatrix}$. For

what values of λ is the gradient method locally con-

vergent when applied to this function? Can you obtain

an estimate for the error at the $(k+1)$st step, i. e.

$\left\| \begin{pmatrix} x_{k+1} \\ y_{k+1} \end{pmatrix} - \begin{pmatrix} x^* \\ y^* \end{pmatrix} \right\|_e$, in terms of the error at the kth step,

$\left\| \begin{pmatrix} x_k \\ y_k \end{pmatrix} - \begin{pmatrix} x^* \\ y^* \end{pmatrix} \right\|$?

2. Let $g : R^1 \rightarrow R^1$ be twice continuously differentiable

and assume

$$x^* = g(x^*),$$

$$\frac{dg}{dx}(x^*) = 0.$$

Show that the iterative method

$$x_{k+1} = g(x_k)$$

yields a sequence $\{x_k\}$ which converges <u>quadratically</u>

to x^*. That is, there is a constant $K > 0$ such that

$$|x_{k+1} - x^*| \leq K|x_k - x^*|^2.$$

Compare this rate of convergence with that of Theorem

8. 2, which is called <u>linear</u> convergence. The result of

this exercise remains true if $g : R^n \to R^n$, $x^* = g(x^*)$,

$\frac{\partial g}{\partial x}(x^*) = 0$ but the proof involves use of the <u>third order</u>

tensor $\frac{\partial^2 g}{\partial x^2}$ which we have not discussed.

3. Let $f : R^1 \to R^1$ be twice continuously differentiable and

assume

$$f(x^*) = 0$$

$$\frac{df}{dx}(x^*) \neq 0 .$$

Prove that Newton's method $x_{k+1} = x_k - \dfrac{f(x_k)}{\dfrac{df}{dx}(x_k)}$

converges quadratically to x^*.

4. Let $f : R^2 \to R^1$ be given by

$$f(x, y) = x^2 + 3xy + 4y^2 .$$

Prepare a careful sketch showing:

(i) The contour curves S_c for $c = 1, 2, 3, 4, 5$.

(ii) The path of steepest descent which passes through

the point $\begin{pmatrix} x_0 \\ y_0 \end{pmatrix} = \begin{pmatrix} \sqrt{5} \\ 0 \end{pmatrix}$. (Also give a formula for

this path.)

(iii) The points $\begin{pmatrix} x_k \\ y_k \end{pmatrix}$, $k = 0, 1, 2, 3, 4, 5$ generated by the gradient method with $\lambda = 0.2$, $\binom{x_0}{y_0} = \binom{\sqrt{5}}{0}$.

5. The function $f : R^2 \to R^1$ defined by

$$f(x, y) = x^2 + 3xy + 4y^2 + \frac{1}{2}\sin(x + 2y)$$

has a relative minimum near $\binom{0}{0}$. Use the gradient

method to find it correctly with a tolerance 1×10^{-3}.

You may wish to use Aitken's Δ^2 process to speed up

the convergence.

6. The Wotzit Company is looking for a location for its

headquarters. From this headquarters executives will

fly by company aircraft to New York, Chicago, Cleveland,

Minneapolis, Atlanta, Miami and New Orleans. Roughly

the same number of trips will be made to each city.

Therefore the company would like to find that location

which minimizes the sum of the distances to these cities.

Use the gradient method to assist the Wotzit Company

in this project. Hint: for coordinates use

x = . 08 x (degrees of longitude)

y = . 1 x (degrees of latitude)

Give your answer in degrees and minutes.

CHAPTER 9. EQUALITY CONSTRAINTS: GRADIENT PROJECTION

In Theorem 5.2 we presented a necessary condition for a function f to assume a minimum at a point x^* in the interior of a set $S \subseteq R^n$, i. e. , $\frac{\partial f}{\partial x}(x^*) = 0$. The techniques presented in Chapters 5 and 8, i. e. , Newton's method and the gradient method, are very much related to this result. It should be noted that the proof of Theorem 5.2 relies heavily on the fact that some neighborhood of x^* is included in the set S.

Let us consider now a problem of a different kind. In R^2 we consider the function

$$f(x, y) = x^2 + y^2$$

and we define a set

$$S = \{ ({}^x_y) \in R^2 \, | \, x + y = 1 \} .$$

The set S is convex and f is strictly convex on S with

$\lim\limits_{\| ({}^x_y) \| \to \infty} f(x, y) = \infty$ so we conclude that there is exactly

one point $({}^{x^*}_{y^*}) \in S$ where f assumes its minimum value in

S. In fact it is quite clear in this example that $({}^{x^*}_{y^*})$ is the

point in S lying closest to the origin, obviously the point

$({}^{\frac{1}{2}}_{\frac{1}{2}})$. The gradient of f at this point is the vector $(1, 1)$

which is not zero. Thus Theorem 5.2 cannot be of much use

in studying problems such as these. We ask the reader to

note, however, that in this example S is a straight line and

the vector $(1, 1)$ is perpendicular to this line.

We are now ready to pose the general problem to be

studied in this chapter. Let D be some open set in R^n and

let $f, g_1, g_2, \ldots, g_m$, $m < n$, be continuously differentiable

functions defined on D. We let S be the subset of D de-

fined by

$$S = \{ x \in D \, | \, g_i(x) = 0, \; i = 1, 2, \ldots, m \}$$

and we seek a point $x^* \in S$ such that

$$f(x^*) \leq f(x), \quad x \in S.$$

The equations $g_i(x) = 0$, $i = 1, 2, \ldots, m$, which define the feasible set S are referred to as equality constraints, the word "equality" distinguishes constraints of this type from those to be studied in Chapter 10.

Our first task will be to study the nature of the set S. Broadly speaking, S is a surface of dimension $n - m$ in the space R^n. Thus the set

$$S = \{(\begin{smallmatrix} x \\ y \\ z \end{smallmatrix}) \in R^3 \mid x^2 + y^2 + z^2 - 1 = 0\}$$

which is described by just one equality constraint (i. e., $m = 1$), is a surface of dimension $3 - 1 = 2$ in R^3, in fact S is a sphere in R^3 in this case. On the other hand

$$\hat{S} = \{(\begin{smallmatrix} x \\ y \\ z \end{smallmatrix}) \in R^3 \mid x^2 + y^2 + z^2 - 1 = 0, \ x + y + z - 1 = 0\},$$

described by two equality constraints (i. e., $m = 2$), is a surface of dimension $3 - 2 = 1$, i. e., a curve, in R^3. The curve $\hat{S}$ is a circle which is the intersection of the sphere S with the plane $x + y + z = 1$. There are some exceptions to this rule. In R^4 the set S described by the three equations

$$w + x + y + z = 0, \qquad w - x + y - z = 0, \qquad w + y = 0$$

is a (plane) surface of dimension 2 rather than 1. The explanation for this lies in the fact that the vectors $(1, 1, 1, 1)$, $(1, -1, 1, -1)$, $(1, 0, 1, 0)$ in $(R^4)'$ are linearly dependent. To ensure that our m constraints, $m < n$, define a surface of dimension $n - m$ in R^n we make the following

ALGEBRAIC CONSTRAINT QUALIFICATION. For each point $\hat{x} \in S = \{x \in R^n \mid g_i(x) = 0, \ i = 1, 2, \ldots, m\}$ the vectors $\dfrac{\partial g_i}{\partial x}(\hat{x})$, $i = 1, 2, \ldots, m$, are linearly independent in $(R^n)'$.

Having made this qualification we can proceed to describe the local geometry of the surface S in some detail. To this end we introduce two important subspaces of R^n.

DEFINITION 9.1. Let $\hat{x} \in S = \{x \in R^n \mid g_i(x) = 0, \ i = 1, 2, \ldots, n\}$. We define $N(\hat{x})$, the _normal space for_ S _at_ $\hat{x}$, to be the subspace of R^n spanned by the vectors $\dfrac{\partial g_i}{\partial x}(\hat{x})'$, $i = 1, 2, \ldots, m$; thus

$$N(\hat{x}) = \{y \in R^n \mid y = \alpha^1 \frac{\partial g_1}{\partial x} (\hat{x})' + \ldots + \alpha^m \frac{\partial g_m}{\partial x} (\hat{x})',$$

$$\alpha^1, \alpha^2, \ldots, \alpha^m \in R^1\}.$$

We define $T(\hat{x})$, <u>the tangent space for</u> S <u>at</u> $\hat{x}$ by

$$T(\hat{x}) = \{z \in R^n \mid y'z = 0, \ y \in N(\hat{x})\}.$$

Referring to Theorem 7.4 and to Problem 2 in the Exercises for Chapter 7 we see that $T(\hat{x}) = N(\hat{x})^\perp$ and $R^n = N(\hat{x}) \oplus T(\hat{x})$. Thus, given any vector $w \in R^n$ there are uniquely determined vectors $y \in N(\hat{x})$, $z \in T(\hat{x})$ such that $w = y + z$.

EXAMPLE. Let $S = \{(\begin{smallmatrix} x^1 \\ x^2 \end{smallmatrix}) \in R^2 \mid (x^1)^2 + (x^2)^2 - 1 = 0\}$ and

consider the point $(\begin{smallmatrix} \hat{x}^1 \\ \hat{x}^2 \end{smallmatrix}) = (\begin{smallmatrix} 3/5 \\ 4/5 \end{smallmatrix}) \in S$. Then

$$N(\begin{smallmatrix} \hat{x}^1 \\ \hat{x}^2 \end{smallmatrix}) = N(\begin{smallmatrix} 3/5 \\ 4/5 \end{smallmatrix}) = \{(\begin{smallmatrix} y^1 \\ y^2 \end{smallmatrix}) \in R^2 \mid (\begin{smallmatrix} y^1 \\ y^2 \end{smallmatrix}) = \alpha(\begin{smallmatrix} 2\hat{x}^1 \\ 2\hat{x}^2 \end{smallmatrix}) = \alpha(\begin{smallmatrix} 6/5 \\ 8/5 \end{smallmatrix})\}$$

$$= \{(\begin{smallmatrix} y^1 \\ y^2 \end{smallmatrix}) \in R^2 \mid 4y^1 = 3y^2\}$$

while

$$T(\begin{smallmatrix} \hat{x}^1 \\ \hat{x}^2 \end{smallmatrix}) = T(\begin{smallmatrix} 3/5 \\ 4/5 \end{smallmatrix}) = \{(\begin{smallmatrix} z^1 \\ z^2 \end{smallmatrix}) \in R^2 \mid 3z^1 + 4z^2 = 0\}.$$

In this case $N(\hat{x}^1_{\hat{x}2})$ and $T(\hat{x}^1_{\hat{x}2})$ are mutually perpendicular

lines through the origin in R^2 as shown in Figure 15.

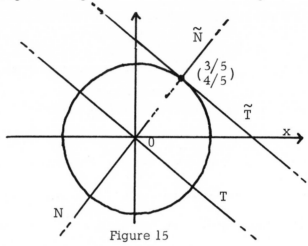

Figure 15

It is often conceptually convenient to think of the

spaces $N(\hat{x})$ and $T(\hat{x})$ as passing through the point $\hat{x}$ rather

than through the origin. Mathematically speaking, what we

are doing is replacing $N(\hat{x})$ and $T(\hat{x})$ by

$$\tilde{N}(\hat{x}) = N(\hat{x}) + \{\hat{x}\} = \{u \in R^n \,|\, u = y + \hat{x},\ y \in N(\hat{x})\},$$

$$\tilde{T}(\hat{x}) = T(\hat{x}) + \{\hat{x}\} = \{v \in R^n \,|\, v = z + \hat{x},\ z \in T(\hat{x})\}.$$

In the example given above it turns out that

$$\tilde{N}\binom{3/5}{4/5} = N\binom{3/5}{4/5}$$

because $(\begin{smallmatrix} 3/5 \\ 4/5 \end{smallmatrix})$ just happens to lie in the space $N(\begin{smallmatrix} 3/5 \\ 4/5 \end{smallmatrix})$ but

$$\widetilde{T}(\begin{smallmatrix} 3/5 \\ 4/5 \end{smallmatrix}) = \{(\begin{smallmatrix} x \\ y \end{smallmatrix}) \in R^2 \mid 3x + 4y = 5\} = T(\begin{smallmatrix} 3/5 \\ 4/5 \end{smallmatrix}) + \{(\begin{smallmatrix} 3/5 \\ 4/5 \end{smallmatrix})\}.$$

In Figure 16 we illustrate the general situation with S a two

dimensional surface in R^3.

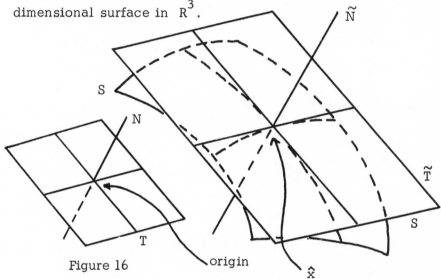

Figure 16

The precise relationship between the tangent space $T(\hat{x})$ and

the surface S is contained in the following two theorems.

THEOREM 9.1. Let $\hat{x}$ be a given point in S. For $x \in S$ let

$w(x) = x - \hat{x}$ and let $y(x)$ and $z(x)$ be the uniquely defined

vectors in $N(\hat{x})$, $T(\hat{x})$, respectively, such that $w(x) = y(x) +$

$z(x)$. Then

$$\lim_{w(x)\ =\ x-\hat{x}\ \to\ 0} \frac{\|y(x)\|}{\|w(x)\|} = 0$$

$$\lim_{w(x)\ =\ x-\hat{x}\ \to\ 0} \frac{\|z(x)\|}{\|w(x)\|} = 1 \ ,$$

i. e. , using the " $\mathcal{O}$ " notation

$$y(x) = \mathcal{O}(w(x)), \ z(x) = w(x) + \mathcal{O}(w(x)), \quad x \to \hat{x}.$$

Proof.   Since $y(x) \in N(\hat{x})$ we can write

$$y(x) = \sum_{j=1}^{m} \alpha^j(x) \frac{\partial g_i}{\partial x}(\hat{x})'.$$

Since $x \in S$ we have $g_i(x) = 0$, $i = 1, 2, \ldots, m$.   Then we have, from the definition of $\dfrac{\partial g_i}{\partial x}(\hat{x})$, the fact that $\hat{x} \in S$ and the identity $w(x) = x - \hat{x}$, that

$$0 = g(x) - g(\hat{x}) = \frac{\partial g_i}{\partial x}(\hat{x})w(x) + \mathcal{O}(w(x)).$$

Since $z(x) \in T(\hat{x})$ and $\dfrac{\partial g_i}{\partial x}(\hat{x}) \in N(\hat{x})$ we have $\dfrac{\partial g_i}{\partial x}(\hat{x})z(x) = 0$

and thus, since $w(x) = y(x) + z(x)$,

$$0 = \frac{\partial g_i}{\partial x}(\hat{x})(y(x)+z(x)) + \mathscr{O}(w(x)) = \frac{\partial g_i}{\partial x}(\hat{x})y(x) + \mathscr{O}(w(x))$$

$$= \frac{\partial g_i}{\partial x}(\hat{x})(\sum_{j=1}^{m} \alpha^j(x)\frac{\partial g_j}{\partial x}(\hat{x})') + \mathscr{O}(w(x)), \quad i = 1, 2, \ldots, m.$$

Thus, for $i = 1, 2, \ldots, m$

$$\sum_{j=1}^{m} (\frac{\partial g_i}{\partial x}(\hat{x})\frac{\partial g_j}{\partial x}(\hat{x})')\alpha^j(x) = \mathscr{O}(w(x)).$$

Letting $G(\hat{x})$ be the symmetric $m \times m$ matrix whose (i,j)th entry is $\frac{\partial g_i}{\partial x}(\hat{x})\frac{\partial g_j}{\partial x}(\hat{x})'$ and letting $\alpha(x)$ be the m-vector with components $\alpha^j(x)$ we have

$$G(\hat{x})\alpha(x) = \mathscr{O}(w(x)).$$

Now $G(\hat{x})$ is not only symmetric, it is also positive definite. To see this we let $\beta = \begin{pmatrix} \beta^1 \\ \vdots \\ \beta^m \end{pmatrix} \epsilon \ R^m$ and note that

$$\beta'G(\hat{x})\beta = \|\sum_{j=1}^{m} \beta^j \frac{\partial g_j}{\partial x}(\hat{x})'\|_e^2 > 0 \quad \text{if} \quad \beta \neq 0$$

since the vectors $\frac{\partial g_j}{\partial x}(\hat{x})'$, $j = 1, 2, \ldots, m$, are linearly independent in R^n by virtue of the algebraic constraint qualification. Thus $G(\hat{x})$ is positive definite and invertible so that

$$\alpha(x) = G(\hat{x})^{-1} \theta(w(x)) = \theta(w(x)).$$

Letting $\Gamma(\hat{x})$ be the $m \times n$ matrix whose rows are the vectors $\dfrac{\partial g_i}{\partial x}(\hat{x})$, $i = 1, 2, \ldots, m$, the expansion of $y(x)$ in terms of these vectors can be written

$$y(x) = \Gamma(\hat{x})'\alpha(x)$$

and the relationship $\alpha(x) = \theta(w(x))$ then gives

$$y(x) = \theta(w(x)).$$

Then

$$z(x) = w(x) - y(x) = w(x) + \theta(w(x))$$

completes the proof of the theorem.

THEOREM 9.2.   Let $\hat{x} \in S$ and let $N(\hat{x})$, $T(\hat{x})$ be, respectively, the normal and tangent spaces for S at $\hat{x}$. There is a positive number ρ such that to each $z \in T(\hat{x})$ with $\|z\| < \rho$ there corresponds a vector $y(z) \in N(\hat{x})$ such that

$$x(z) = \hat{x} + z + y(z) \in S.$$

Moreover, $y(z) = \theta(z)$ as $z \to 0$ in $T(\hat{x})$.

Although we have in Theorem 8.1 a result which would enable us to prove this theorem, we will not present such a proof here because it would lead us to far astray from the line of thought which we are trying to develop. We will indicate later how $y(z)$ would be found in practice. In Exercise 1 at the end of this chapter we will outline the proof of this theorem so that the interested student may develop his own proof.

Theorems 9.1 and 9.2 together show that in a neighborhood of the point $\hat{x} \in S$ there is a one to one correspondence between points $\hat{x} + z$, $z \in T(\hat{x})$, and points $x \in S$ with the difference $x - (\hat{x} + z)$ being a point $y \in N(\hat{x})$. This relationship is illustrated in Figure 17.

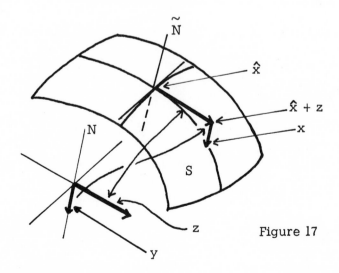

Figure 17

Theorem 9.2 is the principle result which we need to prove the Lagrange Multiplier Theorem, the theoretical foundation for the methods discussed in this chapter.

THEOREM 9.3 (Lagrange Multiplier Theorem). Let $x^* \in S$ be such that for some $\rho_0 > 0$

$$f(x^*) \leq f(x), \quad x \in S, \quad \|x - x^*\| < \rho_0$$

so that x^* is a local minimum for f in S. Then, assuming that $f : R^n \to R^1$ is continuously differentiable near x^*, we have

$$\frac{\partial f}{\partial x}(x^*)' \in N(x^*).$$

Thus there are uniquely defined real numbers $\mu^1, \mu^2, \ldots, \mu^m$, called Lagrange Multipliers, such that

$$\frac{\partial f}{\partial x}(x^*) = \sum_{i=1}^{m} \mu^i \frac{\partial g_i}{\partial x}(x^*).$$

Proof. Since $\frac{\partial f}{\partial x}(x^*)' \in R^n = N(x^*) \oplus T(x^*)$ there are uniquely defined vectors $\tilde{y} \in N(x^*)$, $\tilde{z} \in T(x^*)$ such that

$$\frac{\partial f}{\partial x}(x^*)' = \tilde{y} + \tilde{z}.$$

Since $\tilde{z} \in T(x^*)$, so does $z(\lambda) = -\lambda \tilde{z}$, $\lambda > 0$. If λ is sufficiently small we shall have $\|z(\lambda)\| < \rho$, where ρ is as in Theorem 9.2. Then there is a vector $y(z(\lambda)) \in N(x^*)$ such that

$$x(z(\lambda)) = x^* + z(\lambda) + y(z(\lambda)) \in S$$

and $y(z(\lambda)) = \mathscr{O}(z(\lambda)) = \mathscr{O}(\lambda)$ as $\lambda \to 0$. Clearly then if we take $\lambda > 0$ but sufficiently small we will have $\|x(z(\lambda)) - x^*\| < \rho_0$ and thus

$$0 \geq f(x^*) - f(x(z(\lambda))) = -\frac{\partial f}{\partial x}(x^*)(x(z(\lambda)) - x^*) + \mathscr{O}(x(z(\lambda)) - x^*)$$

$$= -(\tilde{y} + \tilde{z})'(-\lambda \tilde{z} + y(z(\lambda))) + \mathscr{O}(-\lambda \tilde{z} + y(z(\lambda)))$$

$$= \lambda \|\tilde{z}\|_e^2 - \tilde{y}'y(z\lambda)) + \mathscr{O}(\lambda) = \lambda \|\tilde{z}\|_e^2 + \mathscr{O}(\lambda)$$

since $y(z(\lambda)) = \mathscr{O}(\lambda)$.

If $\|\tilde{z}\|_e$ were different from zero, then $\lambda \|\tilde{z}\|_e^2 + \mathscr{O}(\lambda)$ would be positive for small positive values of λ, contradicting the inequality $0 \geq f(x^*) - f(x(\lambda))$. We conclude that $\|\tilde{z}\|_e = 0$, i.e., $\tilde{z} = 0$. Thus

$$\frac{\partial f}{\partial x}(x^*)' = \tilde{y} \in N(x^*)$$

as claimed. Since the linearly independent vectors $\frac{\partial g_i}{\partial x}(x^*)'$

span the space $N(x^*)$, $\tilde{y}$ must be uniquely expressible as a

linear combination of these vectors, i. e. ,

$$\frac{\partial f}{\partial x}(x^*)' = \tilde{y} = \sum_{i=1}^{m} \mu^i \frac{\partial g_i}{\partial x}(x^*)$$

and this completes the proof.

The reader should note how this result agrees with the

example presented in the second paragraph of this chapter.

Theorem 9. 3 says, in effect, that at x^* the gradient of f

must lie perpendicular to the surface S, by which we mean

that it must be perpendicular to all vectors in the tangent

space for S at x^*.

It should be noted that the Lagrange multiplier theorem

does not distinguish between maxima of f and minima of f

in the set S. The condition for f to have a local maximum

at x^* is that $-f$ should have a minimum there which gives

$$-\frac{\partial f}{\partial x}(x^*)' \in N(x^*).$$

But, since $N(x^*)$ is a subspace of R^n, $-\frac{\partial f}{\partial x}(x^*)' \in N(x^*)$ if

and only if $\frac{\partial f}{\partial x}(x^*)' \in N(x^*)$.

Just as in the case for unconstrained problems, where the condition was $\frac{\partial f}{\partial x}(x^*) = 0$, the condition of Theorem 9.3 is a <u>necessary</u> condition rather than a sufficient condition. The condition $\frac{\partial f}{\partial x}(x^*)' \in N(x^*)$ is not enough to guarantee that f has either a local maximum or a local minimum at x^*. An example is provided by

$$f(x,y) = y - x^3,$$

$$S = \{\begin{pmatrix} x \\ y \end{pmatrix} \in R^2 \,|\, y = 0\}, \quad (\text{i. e. }, \; g(x,y) = y)$$

$$(x^*, y^*) = (0,0).$$

Nevertheless, with all these qualifications, Theorem 9.3 provides us with a result which is useful for the determination of x^* in many cases. Sometimes we can use the result fairly directly.

EXAMPLE. Let us try to minimize

$$f(x,y) = x^2 + y^2$$

subject to the single constraint

$$g(x,y) = y + (x + 1)^2 = 0.$$

The gradient of the objective function is the row vector $(2x, 2y)$ whereas the gradient of g is the row vector $(2x + 2, 1)$. Using Theorem 9.3 we see that if (x^*, y^*) solves the problem then there must be a real number μ such that

$$(2x^*, 2y^*) = \mu(2x^* + 2, 1).$$

Combining this with $y^* + (x^* + 1)^2 = 0$ we get three equations in three unknowns

$$2x^* = 2\mu x^* + 2\mu,$$

$$2y^* = \mu,$$

$$y^* + (x^* + 1)^2 = 0.$$

We solve the first two equations for x^*, y^* in terms of μ:

$$x^* = \frac{\mu}{1 - \mu},$$

$$y^* = \frac{\mu}{2},$$

and we substitute in the third equation to get

$$\frac{\mu}{2} + (\frac{\mu}{1-\mu} + 1)^2 = \frac{\mu}{2} + (\frac{1}{1-\mu})^2 = 0$$

or

$$\mu(1-\mu)^2 + 2 = \mu - 2\mu^2 + \mu^3 + 2 = 0.$$

The only real root of this cubic equation $\mu = -.698$, is readily obtained using Newton's method.   Then

$$x^* = \frac{\mu}{1-\mu} = \frac{-.698}{1.698} = -.411$$

$$y^* = \frac{\mu}{2} = \frac{-.698}{2} = -.349.$$

Unfortunately, it is rarely true that a practical constrained optimization problem will be as easily treated as was this "textbook" problem.   We again need a general method for approximating solutions.

What we will develop is a method very similar to the gradient method of the previous chapter - in fact it is called the gradient projection method for problems with equality constraints.   Let us suppose that we are able to find some point x_0 which lies on the surface S.   What we wish to do now is to move along the surface S in such a way as to decrease the values of the function f. In general we cannot set

$x_1 = x_0 - \lambda \frac{\partial f}{\partial x}(x_0)'$ because the point x_1 thus obtained will

not lie on S.

Let $N(x_0)$ and $T(x_0)$ be the normal and tangent

spaces to S at x_0. Then there are uniquely determined vec-

tors $y_0 \in N(x_0)$, $z_0 \in T(x_0)$ such that

$$\frac{\partial f}{\partial x}(x_0)' = y_0 + z_0 .$$

Thus y_0 is the component of $\frac{\partial f}{\partial x}(x_0)'$ which points away

from the surface S while z_0 is the component which lies

tangential to S at x_0. Let us put

$$\hat{x}_1 = x_0 - \lambda z_0 , \qquad \lambda > 0 ,$$

and investigate the value of

$$f(\hat{x}_1) = f(x_0 - \lambda z_0) = f(x_0) - \lambda \frac{\partial f}{\partial x}(x_0)z_0 + \mathscr{O}(\lambda z_0)$$

$$= f(x_0) - \lambda (y_0' + z_0')z_0 + \mathscr{O}(\lambda) .$$

Since $y_0 \in N(x_0)$, $z_0 \in T(x_0)$ we have $y_0'z_0 = 0$ from the

definition of $T(x_0)$. Thus

$$f(\hat{x}_1) = f(x_0) - \lambda z_0'z_0 + \mathscr{O}(\lambda) = f(x_0) - \lambda \|z_0\|_e^2 + \mathscr{O}(\lambda) .$$

For small values of $\lambda > 0$ we shall have $f(\hat{x}_1) < f(x_0)$ unless $z_0 = 0$, i.e., unless $\frac{\partial f}{\partial x}(x_0)' = y_0 \in N(x_0)$, which we would normally take as an indication that x_0 might well be a solution to our problem already. We see, therefore, that if the Lagrange multiplier condition does not hold at x_0, we can decrease f by moving in the direction of $-z_0$.

The reader is probably wondering by now just how we can compute z_0, since all we have done up to now is to assert its existence. Since $y_0 \in N(x_0)$, which is spanned by the vectors $\frac{\partial g_i}{\partial x}(x_0)'$, $i = 1, 2, \ldots, m$, we have

$$\frac{\partial f}{\partial x}(x_0)' = y_0 + z_0 = \sum_{i=1}^{m} \alpha_0^i \frac{\partial g_i}{\partial x}(x_0)' + z_0$$

for some real coefficients α_0^i, as yet unknown. Since $z_0 \in T(x_0)$ and $\frac{\partial g_j}{\partial x}(x_0)' \in N(x_0)$, $j = 1, 2, \ldots, m$, we must have

$$0 = \frac{\partial g_j}{\partial x}(x_0) z_0 = \frac{\partial g_j}{\partial x}(x_0)[\frac{\partial f}{\partial x}(x_0)' - \sum_{i=1}^{m} \alpha_0^i \frac{\partial g_i}{\partial x}(x_0)']$$

which gives, for $j = 1, 2, \ldots, m$, the equation

$$\sum_{i=1}^{m} (\frac{\partial g_j}{\partial x}(x_0) \frac{\partial g_i}{\partial x}(x_0)') \alpha_0^i = \frac{\partial g_j}{\partial x}(x_0) \frac{\partial f}{\partial x}(x_0)'.$$

Recalling that $G(x_0)$ is the symmetric positive definite $m \times m$ matrix with entries $\frac{\partial g_i}{\partial x}(x_0) \frac{\partial g_j}{\partial x}(x_0)'$ and that $\Gamma(x_0)$ is the $m \times n$ matrix with rows $\frac{\partial g_i}{\partial x}(x_0)$, the m equations of the preceding formula can be summarized by the vector equation

$$G(x_0)\alpha_0 = \Gamma(x_0)\frac{\partial f}{\partial x}(x_0)'$$

whence

$$\alpha_0 = G(x_0)^{-1}\Gamma(x_0)\frac{\partial f}{\partial x}(x_0)' .$$

Now the relationship

$$z_0 = \frac{\partial f}{\partial x}(x_0)' - \sum_{i=1}^{m} \alpha_0^i \frac{\partial g_i}{\partial x}(x_0)'$$

can also be written as

$$z_0 = \frac{\partial f}{\partial x}(x_0)' - \Gamma(x_0)'\alpha_0 .$$

Combining this with the value of α_0 already obtained we have

$$z_0 = \frac{\partial f}{\partial x}(x_0)' - \Gamma(x_0)'G(x_0)^{-1}\Gamma(x_0)\frac{\partial f}{\partial x}(x_0)'$$

$$= [I - \Gamma(x_0)'G(x_0)^{-1}\Gamma(x_0)]\frac{\partial f}{\partial x}(x_0)' .$$

Our formula for $\hat{x}_1$ is then

$$\hat{x}_1 = x_0 - \lambda [I - \Gamma(x_0)' G(x_0)^{-1} \Gamma(x_0)] \frac{\partial f}{\partial x}(x_0)'$$

$$= x_0 - \lambda Q(x_0) \frac{\partial f}{\partial x}(x_0)', \quad Q(x_0) = I - \Gamma(x_0)' G(x_0)^{-1} \Gamma(x_0).$$

We call $Q(x_0)$ the <u>projection matrix onto</u> $T(x_0)$. In fact, if w is any vector in R^n and we decompose w into components in $N(x_0)$ and $T(x_0)$, $w = y + z$, then $z = Q(x_0)w$.

The determination of $\hat{x}_1$ as described above does not really satisfy our objective of moving along the surface S. Although $\hat{x}_1 - x_0 = -\lambda z_0$ is tangent to S at x_0 the point $\hat{x}_1$ is in general slightly removed from S. Only if all of the constraint functions $g_i(x)$ are linear can we be sure that $\hat{x}_1$ belongs to S. In other cases we need a further correction step which we will describe now.

If in Theorem 9.2 we replace $\hat{x}$ by the vector x_0 above and replace z by $-\lambda z_0$, then that theorem tells us that if $\lambda > 0$ is sufficiently small there is a vector $y_0 \epsilon N(x_0)$ such that

$$x_1 = x_0 - \lambda z_0 + y_0 = \hat{x}_1 + y_0 \epsilon S,$$

that is

$$g_i(x_1) = g_i(\hat{x}_1 + y_0) = 0 , \quad i = 1, 2, \ldots, m .$$

For convenience we let g be the vector valued function with components g_i, i. e. ,

$$g(x) = \begin{pmatrix} g_1(x) \\ \vdots \\ g_m(x) \end{pmatrix} .$$

Then the m equations above can be written

$$g(x_1) = g(\hat{x}_1 + y_0) = 0 .$$

The problem is to find y_0. Since $y_0 \in N(x_0)$ there are real numbers $\alpha_0^1, \alpha_0^2, \ldots, \alpha_0^m$ such that

$$y_0 = \sum_{j=1}^{m} \alpha^j \frac{\partial g_j}{\partial x}(x_0)' ,$$

i. e. ,

$$y_0 = \Gamma(x_0)' \alpha_0 , \quad \alpha_0 = \begin{pmatrix} \alpha_0^1 \\ \vdots \\ \alpha_0^m \end{pmatrix} .$$

So we may now ask for the value of α_0 which solves the equation in α :

$$\tilde{g}(\alpha) = g(\hat{x}_1 + \Gamma(x_0)'\alpha) = 0.$$

Newton's method for solving equations of this type was discussed in Chapter 5. Starting with $\alpha_{00} = 0$, which is a logical choice since we expect y_0 to be a small vector, we generate a sequence $\{\alpha_{0,k}\}$ by

$$\alpha_{0,k+1} = \alpha_{0,k} - (\frac{\partial \tilde{g}}{\partial \alpha}(\alpha_{0,k}))^{-1}\tilde{g}(\alpha_{0,k}).$$

Now

$$\frac{\partial \tilde{g}(\alpha)}{\partial \alpha} = \frac{\partial}{\partial \alpha}(g(\hat{x}_1 + \Gamma(x_0)'\alpha) = \frac{\partial g}{\partial x}(\hat{x}_1 + \Gamma(x_0)'\alpha)\Gamma(x_0)'$$

and we see that

$$\alpha_{0,k+1} = \alpha_{0,k} - (\frac{\partial g}{\partial x}(\hat{x}_1 + \Gamma(x_0)'\alpha_{0,k})\Gamma(x_0)')^{-1}g(\hat{x}_1 + \Gamma(x_0)'\alpha_{0,k}).$$

It can be shown that the sequence $\{\alpha_{0,k}\}$ generated by this method converges to a solution α_0 of $\tilde{g}(\alpha) = 0$ if $\hat{x}_1$ is sufficiently close to x_0, i.e., if λ is sufficiently small.

In practice it would be quite wasteful of time to use Newton's method just as described above because we would have to recalculate $\frac{\partial g}{\partial x}(\hat{x}_1 + \Gamma(x_0)'\alpha_{0,k})$ for each new value of k and invert the matrix. We note that $\frac{\partial g}{\partial x}(x_0) = \Gamma(x_0)$.

Since the point $\hat{x}_1 + \Gamma(x_0)' \alpha_{0,k}$ lies near x_0 we may ex-

pect that $\Gamma(x_0)$ is a good approximation to $\frac{\partial g}{\partial x}(\hat{x}_1 + \Gamma(x_0)' \alpha_{0,k})$

and hence that $G(x_0)^{-1} = (\Gamma(x_0)\Gamma(x_0)')^{-1}$ is a good approxi-

mation to $(\frac{\partial g}{\partial x}(\hat{x}_1 + \Gamma(x_0)' \alpha_{0,k})\Gamma(x_0)')^{-1}$. It is therefore

reasonable to generate the sequence $\{\alpha_0, k\}$ by the simpli-

fied formula

$$\alpha_{0,k+1} = \alpha_{0,k} - G(x_0)^{-1} g(\hat{x}_1 + \Gamma(x_0)' \alpha_{0,k}).$$

Referring to Chapter 8 we note that what we have done is

exactly analogous to the manner in which we obtained the

modified gradient method $x_{k+1} = x_k - P\frac{\partial f}{\partial x}(x_k)'$, where P

was an approximation to the inverse Hessian matrix. In the

formula above $G(x_0)^{-1}$ is an approximation to $(\frac{\partial \tilde{g}}{\partial \alpha}(\alpha_{0,k}))^{-1}$.

We now give a summary of the complete gradient pro-

jection method.

SUMMARY

(i)   Find $x_0 \in S$, i.e., x_0 such that $g_i(x_0) = 0$,

        $i = 1, 2, \ldots, m$ $(g(x_0) = 0)$;

(ii)   Given a point $x_k \in S$, calculate $\Gamma(x_k) = \frac{\partial g}{\partial x}(x_k)$,

$$G(x_k)^{-1} = (\Gamma(x_k)\Gamma(x_k)')^{-1} \text{ and } Q(x_k) = I - \Gamma(x_k)'G(x_k)^{-1}\Gamma(x_k);$$

(iii) Put $\hat{x}_{k+1} = x_k - \lambda Q(x_k)\dfrac{\partial f}{\partial x}(x_k)'$;

(iv) Obtain a new point $x_{k+1} = \hat{x}_{k+1} + \Gamma(x_k)'\alpha_k \in S$ by

solving $g(\hat{x}_{k+1} + \Gamma(x_k)'\alpha_k) = 0$. The solution α_k is

given by $\alpha_k = \lim\limits_{\ell \to \infty} \alpha_{k,\ell}$, where the $\alpha_{k,\ell}$ are gener-

ated by

$$\alpha_{k,0} = 0$$

$$\alpha_{k,\ell+1} = \alpha_k, \quad -G(x_k)^{-1}g(\hat{x}_{k+1} + \Gamma(x_k)'\alpha_{k,\ell}) \ .$$

(v) Go to (ii) and continue the complete process until the

Lagrange multiplier condition is satisfied at x_{k+1} with-

in appropriate tolerances.

Just as with the ordinary gradient method of Chapter 8,

the choice of λ is often more of an art than a science. One

can obtain conditions which are generalizations of the condi-

tion $0 < \lambda < \dfrac{2}{\lambda_1}$ of Theorem 8. 3. These are of little use in

practice since the point x^* which is the solution of our

problem is not available.

The correction procedure (iv) need not always be per-

formed each time. In many cases it might only be performed

after a _series_ of steps in which we have put $x_{k+1} = \hat{x}_{k+1}$.

To see why this should be so we note that the formula

$$x_{k+1} = \hat{x}_{k+1} = x_k - \lambda [I - \Gamma(x_k)' G(x_k)^{-1} \Gamma(x_k)] \frac{\partial f}{\partial x}(x_k)'$$

can be regarded as Euler's method for approximating solutions

of the differential equation

$$\frac{dx}{dt} = -Q(x) \frac{\partial f}{\partial x}(x)' = -[I - \Gamma(x)' G(x)^{-1} \Gamma(x)] \frac{\partial f}{\partial x}(x)'.$$

Now we can easily show that a solution of this differential

equation which starts out on the surface S (i. e. , $g(x(0)) = 0$)

always remains on S (i. e. , $g(x(t)) \equiv 0$, $t \geq 0$). For we have,

if $x(t)$ satisfies the differential equation above,

$$\frac{d}{dt}(g(x(t))) = \frac{\partial g}{\partial x}(x(t)) \frac{dx}{dt}$$

$$= -\frac{\partial g}{\partial x}(x(t))[I - \Gamma(x(t))' G(x(t))^{-1}\Gamma(x(t))] \frac{\partial f}{\partial x}(x(t))'$$

$$= [-\Gamma(x(t)) - \Gamma(x(t))\Gamma(x(t))' G(x(t))^{-1}\Gamma(x(t))] \frac{\partial f}{\partial x}(x(t))^{-1}$$

$$= [-\Gamma(x(t)) - G(x(t))G(x(t))^{-1}\Gamma(x(t))] \frac{\partial f}{\partial x}(x(t))^{-1}$$

$$= [-\Gamma(x(t)) - I\Gamma(x(t))] \frac{\partial f}{\partial x}(x(t))^{-1} \equiv 0.$$

Thus the solution $x(t)$ remains on the surface S.

Now the points x_k remain near the points $x(k\lambda)$ and thus

lie near S if λ is small. (cf. Theorem 8. 4). In Figure 18

we show the relationship of the points x_k to the solution

$x(t)$ in the case where the corrections (iv) are made at each

step and in the case where they are made at each fifth step.

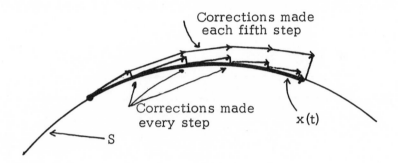

Figure 18

The solution $x(t)$ of $\dfrac{dx}{dt} = - Q(x)\dfrac{\partial f}{\partial x}(x)'$ is a <u>con-</u>

<u>strained path of steepest descent</u> for the function f. It is a

path of steepest descent for the function f restricted to the

surface S. Using this concept one could develop criteria

for the global convergence of the gradient projection method

developed here just as we did for the ordinary gradient method

in Chapter 8.

We will now illustrate the use of the gradient projection

technique in a "space age" example.

EXAMPLE.   We consider a rocket whose shape is that of a

cylinder as shown in Figure 19.   The total mass of the rocket

is 10 units and the payload mass is 1 unit.   The remaining 9

units consists of 6 units of fuel and 3 units of supporting

structure.

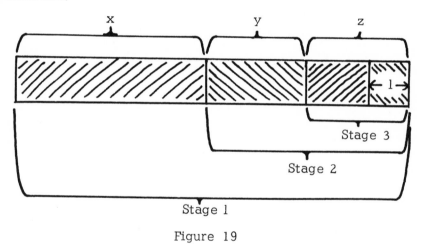

Figure 19

It is known experimentally that when this fuel is used

in a rocket of this type one obtains a velocity increment (in

meters per second, let us say) given by

$$\Delta v = \frac{2000u}{m} \text{ meters per second.}$$

Here u is the number of unit masses of fuel burned and m

is the _initial_ mass of the rocket, including the fuel.

The rocket is to be fired in three stages in a zero-gravity environment. The objective is to impart maximum velocity to the third, or payload, stage.

The cylinder which forms the body of the rocket is 10 units long. Staging is accomplished by detaching a segment of the cylinder after the fuel in that segment has been burned. Thus the rocket cylinder consists of three segments whose lengths we shall call x, y and z as shown in Figure 19. The segment of length z includes the payload which has unit length. The first constraint on the variables x, y and z is clearly

$x + y + z = 10.$

To simplify control procedures during the firing of the first and second stages it has been decided that the ratio of the length (mass) of the third stage to the length (mass) of the second stage should be the same as the ratio of the length of the second to the length of the first stage. Thus we have an additional constraint

$$\frac{z}{y+z} = \frac{y+z}{10} \quad \text{or} \quad (y+z)^2 = 10z.$$

The first stage has mass 10 units and the fuel burned is $\frac{2}{3}x$ units. Thus the initial velocity increment is

$$\Delta v_1 = \frac{2000\,(\frac{2}{3}\,x)}{10} \quad \text{meters per second.}$$

The second stage has mass $y + z$ units and the fuel burned is $\frac{2}{3}y$ units. Thus the second velocity increment is

$$\Delta v_2 = \frac{2000\,(\frac{2}{3}x)}{y+z} \quad \text{meters per second.}$$

The third stage has mass z units and the fuel burned is $\frac{2}{3}(z-1)$ units. Thus the third and final velocity increment is

$$\Delta v_3 = \frac{2000\,(\frac{2}{3}\,(z-1))}{z} = 2000\,(\frac{2}{3}\,(1-\frac{1}{z})) \quad \begin{array}{c}\text{meters per}\\ \text{second.}\end{array}$$

The total velocity increment is

$$\Delta v = 2000\,(\frac{2}{3}[\frac{x}{10}+\frac{y}{y+z}+1-\frac{1}{z}]) \quad \text{meters per second.}$$

The factor $2000\,(\frac{2}{3})$ and the additive constant 1 are clearly irrelevant to the maximization problem. Thus our problem becomes

$$\text{maximize } \frac{x}{10} + \frac{y}{y+z} - \frac{1}{z} = f(x, y, z)$$

subject to the constraints

$$g_1(x, y, z) = x + y + z - 10 = 0,$$

$$g_2(x, y, z) = (y + z)^2 - 10z = 0.$$

Perhaps the easiest way to get an initial point (x_0, y_0, z_0) which lies on the constraint surface is to take

$$x_0 = 5, \quad y_0 = z_0 = 2.5.$$

Since we are maximizing f, which is the same as minimizing $-f$, the gradient projection formula for passing from a point (x_k, y_k, z_k) to a new point $(\hat{x}_{k+1}, \hat{y}_{k+1}, \hat{z}_{k+1})$ is

$$\begin{pmatrix} \hat{x}_{k+1} \\ \hat{y}_{k+1} \\ \hat{z}_{k+1} \end{pmatrix} = \begin{pmatrix} x_k \\ y_k \\ z_k \end{pmatrix} + \lambda Q(x_k, y_k, z_k) \begin{pmatrix} \frac{\partial f}{\partial x}(x_k, y_k, z_k) \\ \frac{\partial f}{\partial y}(x_k, y_k, z_k) \\ \frac{\partial f}{\partial z}(x_k, y_k, z_k) \end{pmatrix}.$$

Here $Q(x_k, y_k, z_k)$ is the projection matrix

$$Q(x_k, y_k, z_k)$$
$$= I - \Gamma(x_k, y_k, z_k)'(\Gamma(x_k, y_k, z_k)\Gamma(x_k, y_k, z_k)')^{-1}\Gamma(x_k, y_k, z_k)$$

with

$$\Gamma(x, y, z) = \begin{pmatrix} \dfrac{\partial g_1}{\partial x} & \dfrac{\partial g_1}{\partial y} & \dfrac{\partial g_1}{\partial z} \\[2mm] \dfrac{\partial g_2}{\partial x} & \dfrac{\partial g_2}{\partial y} & \dfrac{\partial g_2}{\partial z} \end{pmatrix}$$

$$= \begin{pmatrix} 1 & 1 & 1 \\[2mm] 0 & 2(y+z) & 2(y+z) - 10 \end{pmatrix} .$$

Also we have

$$\begin{pmatrix} \dfrac{\partial f}{\partial x} \\[2mm] \dfrac{\partial f}{\partial y} \\[2mm] \dfrac{\partial f}{\partial z} \end{pmatrix} = \begin{pmatrix} \dfrac{1}{10} \\[2mm] \dfrac{z}{(y+z)^2} \\[2mm] \dfrac{1}{z^2} - \dfrac{y}{(y+z)^2} \end{pmatrix} .$$

We will try $\lambda = 5$ in this case because the partial derivatives of f will be fairly small. Also, we will use the above formula repeatedly, making corrections to return to the constraint surface only if the discrepancies between $g_1(\hat{x}_k, \hat{y}_k, \hat{z}_k)$ become "noticeable". (In a computer program one would use some criteria such as $|g_i(\hat{x}_k, \hat{y}_k, \hat{z}_k)| > 10^{-r}$, e. g., where $i = 1$ or 2 and r is some positive integer.)

We start with $x_0 = 5$, $y_0 = z_0 = 2.5$. We will describe the first step in some detail. First we compute the
gradient of $f(x, y, z) = \dfrac{x}{10} + \dfrac{y}{y+z} - \dfrac{1}{z}$ at the initial point:

$$
\begin{pmatrix}
\dfrac{\partial f}{\partial x}(x_0, y_0, z_0) \\[2ex]
\dfrac{\partial f}{\partial y}(x_0, y_0, z_0) \\[2ex]
\dfrac{\partial f}{\partial z}(x_0, y_0, z_0)
\end{pmatrix}
=
\begin{pmatrix}
\dfrac{1}{10} \\[2ex]
\dfrac{2.5}{(2.5+2.5)^2} \\[2ex]
\dfrac{1}{(2.5)^2} - \dfrac{2.5}{(2.5+2.5)^2}
\end{pmatrix}
=
\begin{pmatrix}
.100 \\[1ex]
.100 \\[1ex]
.060
\end{pmatrix}.
$$

We compute $2(y_0 + z_0) = 2(2.5 + 2.5) = 10$ and then

$$
G(x_0, y_0, z_0) = (\Gamma(x_0, y_0, z_0)\Gamma(x_0, y_0, z_0)')^{-1}
$$

$$
= \left[
\begin{pmatrix} 1 & 1 & 1 \\ 0 & 10 & 0 \end{pmatrix}
\begin{pmatrix} 1 & 0 \\ 1 & 10 \\ 1 & 0 \end{pmatrix}
\right]^{-1}
$$

$$
= \begin{pmatrix} 3 & 10 \\ 10 & 100 \end{pmatrix}^{-1}
= \begin{pmatrix} .5 & -.05 \\ -.05 & .015 \end{pmatrix}.
$$

Then the projection matrix is

$Q(x_0, y_0, z_0)$

$$= \begin{pmatrix} 1 & 0 & 0 \\ 0 & 1 & 0 \\ 0 & 0 & 1 \end{pmatrix} - \begin{pmatrix} 1 & 0 \\ 1 & 10 \\ 1 & 0 \end{pmatrix} \begin{pmatrix} .5 & -.05 \\ -.05 & .015 \end{pmatrix} \begin{pmatrix} 1 & 1 & 1 \\ 0 & 10 & 0 \end{pmatrix}$$

$$= \begin{pmatrix} .5 & 0 & -.5 \\ 0 & 0 & 0 \\ -.5 & 0 & .5 \end{pmatrix} .$$

A new point is obtained by adding five times the projection matrix times the gradient of the objective to the initial point:

$$\begin{pmatrix} \hat{x}_1 \\ \hat{x}_2 \\ \hat{x}_3 \end{pmatrix} = \begin{pmatrix} 5.0 \\ 2.5 \\ 2.5 \end{pmatrix} + 5 \begin{pmatrix} .5 & 0 & -.5 \\ 0 & 0 & 0 \\ -.5 & 0 & .5 \end{pmatrix} \begin{pmatrix} .10 \\ .10 \\ .06 \end{pmatrix} = \begin{pmatrix} 5.1 \\ 2.5 \\ 2.4 \end{pmatrix} .$$

We have

$$g_1(\hat{x}_1, \hat{y}_1, \hat{z}_1) = 5.1 + 2.5 + 2.4 - 10 = 0 ,$$

$$g_2(\hat{x}_1, \hat{y}_1, \hat{z}_1) = (2.5 + 2.4)^2 - 10(2.4) = 24.01 - 24.00 = .01.$$

Thus we do not use the correction procedure during this step.

The second step proceeds just like the first:

$$
\begin{pmatrix} \dfrac{\partial f}{\partial x}(\hat{x}_1, \hat{y}_1, \hat{z}_1) \\[2mm] \dfrac{\partial f}{\partial y}(\hat{x}_1, \hat{y}_1, \hat{z}_1) \\[2mm] \dfrac{\partial f}{\partial z}(\hat{x}_1, \hat{y}_1, \hat{z}_1) \end{pmatrix} = \begin{pmatrix} \dfrac{1}{10} \\[2mm] \dfrac{1}{(2.5+2.4)^2} \\[2mm] \dfrac{1}{(2.4)^2} - \dfrac{2.5}{(2.5+2.4)^2} \end{pmatrix} = \begin{pmatrix} .100 \\[2mm] .100 \\[2mm] .070 \end{pmatrix},
$$

$$
G(\hat{x}_1, \hat{y}_1, \hat{z}_1) = \left[\begin{pmatrix} 1 & 1 & 1 \\ 0 & 9.8 & -.2 \end{pmatrix} \begin{pmatrix} 1 & 0 \\ 1 & 9.8 \\ 1 & -.2 \end{pmatrix} \right]^{-1} = \begin{pmatrix} .491 & -.049 \\ -.049 & .015 \end{pmatrix},
$$

$$
Q(\hat{x}_1, \hat{y}_1, \hat{z}_1)
$$

$$
= \left[\begin{pmatrix} 1 & 0 & 0 \\ 0 & 1 & 0 \\ 0 & 0 & 1 \end{pmatrix} - \begin{pmatrix} 1 & 0 \\ 1 & 9.8 \\ 1 & -2 \end{pmatrix} \begin{pmatrix} .491 & -.049 \\ -.049 & .015 \end{pmatrix} \begin{pmatrix} 1 & 1 & 1 \\ 0 & 9.8 & -2 \end{pmatrix} \right]
$$

$$
= \begin{pmatrix} .509 & -.011 & -.501 \\ -.011 & .028 & .009 \\ -.501 & .009 & .489 \end{pmatrix},
$$

$$
\begin{pmatrix} \hat{x}_2 \\ \hat{y}_2 \\ \hat{z}_2 \end{pmatrix} = \begin{pmatrix} 5.1 \\ 2.5 \\ 2.4 \end{pmatrix} + 5 \begin{pmatrix} .509 & -.011 & -.501 \\ -.011 & .028 & .009 \\ -.501 & .009 & .489 \end{pmatrix} \begin{pmatrix} .1 \\ .1 \\ .07 \end{pmatrix}
$$

$$
= \begin{pmatrix} 5.18 \\ 2.52 \\ 2.32 \end{pmatrix},
$$

$$g_1(\hat{x}_2, \hat{y}_2, \hat{z}_2) = 5.18 + 2.52 + 2.32 - 10.00 = .02 ,$$

$$g_2(\hat{x}_2, \hat{y}_2, \hat{z}_2) = (2.52+2.32)^2 - 10(2.32) = 23.43 - 23.2 = .23.$$

Again we will not employ the correction step.

We continue in this manner for a total of six steps, thereby obtaining the following table:

Table 1

k	$\hat{x}_k$	$\hat{y}_k$	$\hat{z}_k$	$\lvert g_1(\hat{x}_k, \hat{y}_k, \hat{z}_k) \rvert$	$\lvert g_2(\hat{x}_k, \hat{y}_k, \hat{z}_k) \rvert$
0	5.00	2.50	2.50	.00	.00
1	5.10	2.50	2.40	.00	.01
2	5.18	2.52	2.32	.02	.23
3	5.25	2.51	2.28	.04	.14
4	5.29	2.50	2.25	.04	.06
5	5.33	2.48	2.21	.02	-.10
6	5.35	2.46	2.19	.00	-.28

The numbers in the last two columns are not too reliable due to round-off errors. However, little accuracy is required since we are using these numbers only to tell us how far we have strayed from the constraint surface.

We will stop at this point and make a correction step.

The correction will be made in the form

$$\begin{pmatrix} x_6 \\ y_6 \\ z_6 \end{pmatrix} = \begin{pmatrix} \hat{\hat{x}}_6 \\ \hat{y}_6 \\ \hat{z}_6 \end{pmatrix} + \Gamma(\hat{\hat{x}}_6, \hat{y}_6, \hat{z}_6) \begin{pmatrix} \alpha^1 \\ \alpha^2 \end{pmatrix}$$

where the coefficients α^1_5, α^2_5 are computed via the recursion

relations

$$\begin{pmatrix} \alpha^1_{\ell+1} \\ \alpha^2_{\ell+1} \end{pmatrix} = \begin{pmatrix} \alpha^1_\ell \\ \alpha^2_\ell \end{pmatrix} - G(\hat{\hat{x}}_6, \hat{y}_6, \hat{z}_6)^{-1} \begin{pmatrix} g_{1,\ell} \\ g_{2,\ell} \end{pmatrix}$$

where $g_{i,\ell} = g_i \left[\begin{pmatrix} \hat{\hat{x}}_6 \\ \hat{y}_6 \\ \hat{z}_6 \end{pmatrix} + \Gamma(\hat{\hat{x}}_6, \hat{y}_6, \hat{z}_6)' \begin{pmatrix} \alpha^1_\ell \\ \alpha^2_\ell \end{pmatrix} \right]$.   This is

slightly different from the formula given before in that we are

evaluating G and Γ at $\begin{pmatrix} \hat{\hat{x}}_6 \\ \hat{y}_6 \\ \hat{z}_6 \end{pmatrix}$ instead of at $\begin{pmatrix} x_5 \\ y_5 \\ z_5 \end{pmatrix}$.

Otherwise there is no difference.

We start with $\begin{pmatrix} \alpha^1_0 \\ \alpha^2_0 \end{pmatrix} = \begin{pmatrix} 0 \\ 0 \end{pmatrix}$. We compute

$$\Gamma(\hat{x}_6, \hat{y}_6, \hat{z}_6) = \begin{pmatrix} 1 & 1 & 1 \\ 0 & 9.3 & -.7 \end{pmatrix} ,$$

$$G(\hat{x}_6, \hat{y}_6, \hat{z}_6)^{-1} = \begin{pmatrix} .464 & -.046 \\ -.046 & .016 \end{pmatrix} .$$

Then we have

$$\begin{pmatrix} \alpha_1^1 \\ \alpha_1^2 \end{pmatrix} = \begin{pmatrix} 0 \\ 0 \end{pmatrix} - \begin{pmatrix} .465 & -.046 \\ -.046 & .016 \end{pmatrix} \begin{pmatrix} .00 \\ -.28 \end{pmatrix} = \begin{pmatrix} -.014 \\ .004 \end{pmatrix} .$$

Only one step of the above procedure is required in this particular case. The corrected values are

$$\begin{pmatrix} x_6 \\ y_6 \\ z_6 \end{pmatrix} = \begin{pmatrix} 5.35 \\ 2.46 \\ 2.19 \end{pmatrix} + \begin{pmatrix} 1 & .0 \\ 1 & 9.3 \\ 1 & -.7 \end{pmatrix} \begin{pmatrix} -.014 \\ .004 \end{pmatrix} = \begin{pmatrix} 5.34 \\ 2.49 \\ 2.17 \end{pmatrix}$$

for which

$$g_1(x_6, y_6, z_6) = x_6 + y_6 + z_6 - 10 = .00 ,$$

$$g_2(x_6, y_6, z_6) = (y_6 + z_6)^2 - 10z_6 = .00$$

and thus we have returned to the constraint surface.

We now continue, as before, using gradient projection and obtain

$$\hat{x}_7 = 5.35, \quad \hat{y}_7 = 2.50, \quad \hat{z}_7 = 2.16.$$

Here we stop. For we compute

$$(\frac{\partial f}{\partial x}(\hat{x}_7,\hat{y}_7,\hat{z}_7), \frac{\partial f}{\partial y}(\hat{x}_7,\hat{y}_7,\hat{z}_7), \frac{\partial f}{\partial z}(\hat{x}_7,\hat{y}_7,\hat{z}_7)) = (.100,.100,.099)$$

whereas

$$(\frac{\partial g_1}{\partial x}(\hat{x}_7,\hat{y}_7,\hat{z}_7), \frac{\partial g_1}{\partial y}(\hat{x}_7,\hat{y}_7,\hat{z}_7), \frac{\partial g_1}{\partial z}(\hat{x}_7,\hat{y}_7,\hat{z}_7)) = (1,1,1).$$

Thus at $(\hat{x}_7,\hat{y}_7,\hat{z}_7)$ the Lagrange multiplier conditions are satisfied approximately with $\mu^1 = .1$, $\mu^2 = 0$. We go through one more correction step to make sure that the constraints are satisfied and get

$$x_7 = 5.35 \qquad y_7 = 2.49 \qquad z_7 = 2.16.$$

Here

$$(\frac{\partial f}{\partial x}(x_7,y_7,z_7), \frac{\partial f}{\partial y}(x_7,y_7,z_7), \frac{\partial f}{\partial z}(x_7,y_7,z_7)) = (.100,.100,.100)$$

and we accept these as the final values.

For the original values $x_0 = 5, y_0 = z_0 = 2.5$ the

total velocity increment is

$$\Delta v = 2000[\,\frac{2}{3}(\frac{x_0}{10} + \frac{y_0}{y_0+z_0} + 1 - \frac{1}{z_0}\,)] = 2135$$

meters per second. For x_7, y_7, z_7 we have

$$\Delta v = 2000[\,\frac{2}{3}(\frac{x_7}{10} + \frac{y_7}{10} + 1 - \frac{1}{z_7}\,)] = 2143$$

meters per second. The increase is not great and the engineers might well decide to stay with the simpler numbers we started with. But, mathematically speaking, we have done our duty.

It is interesting to note that the numbers x_7, y_7, z_7 have the property

$$\frac{x_7}{10} = \frac{y_7}{y_7 + z_7} = \frac{z_7 - 1}{z_7}$$

which means that in all three stages the ratio of the fuel being burned to the mass which it is propelling remains constant. Also interesting is the fact that $\mu^2 = 0$. This means that we could have omitted the constraint $g_2(x, y, z) = 0$ and we still would have obtained the same answer.

EXERCISES. CHAPTER 9

1. The vector $y(z)$ of Theorem 9.2 is found by putting

$y(z) = \Gamma(x)'\alpha(z)$, $\alpha(z) \in R^m$. The vector $\alpha(z)$ is found

by solving

$$g(\hat{x} + z + \Gamma(\hat{x})'\alpha) = 0, \qquad g(x) = \begin{pmatrix} g_1(x) \\ g_2(x) \\ \vdots \\ g_m(x) \end{pmatrix},$$

a system of m equations in m unknowns. Noting that

$\frac{\partial g}{\partial x}(\hat{x}) = \Gamma(\hat{x})$, $\Gamma(\hat{x})z = 0$, $g(\hat{x}) = 0$ and putting

$x(\mu) = \hat{x} + \mu(z + \Gamma(\hat{x})'\alpha)$ we have

$$g(\hat{x} + z + \Gamma(\hat{x})'\alpha) = \int_0^1 \frac{\partial g}{\partial x}(\hat{x} + \mu(z + \Gamma(\hat{x})'\alpha))(z + \Gamma(\hat{x})'\alpha)d\mu$$

$$= \Gamma(\hat{x})\Gamma(\hat{x})'\alpha + \int_0^1 [\frac{\partial g}{\partial x}(\hat{x} + \mu(z + \Gamma(\hat{x})'\alpha)) - \frac{\partial g}{\partial x}(\hat{x})](z + \Gamma(\hat{x})'\alpha)d\mu$$

$$= 0.$$

Recalling that $\Gamma(\hat{x})\Gamma(\hat{x})' = G(x)$ is assumed invertible we

have

$$\alpha = -G(\hat{x})^{-1} \int_0^1 [\frac{\partial g}{\partial x}(\hat{x} + \mu(z + \Gamma(\hat{x})'\alpha)) - \frac{\partial g}{\partial x}(\hat{x})](z + \Gamma(\hat{x})'\alpha)d\mu$$

$$= \tilde{g}(\alpha).$$

Assuming $\frac{\partial g}{\partial x} = \Gamma(x)$ is continuous near $\hat{x}$, show that

if $z \in T(\hat{x})$ and $\|z\|$ is small then there is a closed

ball $\overline{N(0, \delta)}$ in R^m with $\tilde{g} : N(0, \delta) \to N(0, \delta)$ and $\tilde{g}$

is a contraction in $\overline{N(0, \delta)}$. Then Theorem 8.1 applies

to show that $\alpha = \tilde{g}(\alpha)$, and hence $g(\hat{x} + z + \Gamma(\hat{x})'\alpha) = 0$,

has a unique solution in $\overline{N(0, \delta)}$.

2. Let $\hat{x}$ be a point on the surface $S = \{x \in R^n | g_i(x) = 0,$
   $i = 1, 2, \ldots, m\}$ where the algebraic constraint qualification holds. Show that if we put

   $$P(\hat{x}) = \Gamma(\hat{x})'(\Gamma(\hat{x})\Gamma(\hat{x})')^{-1}\Gamma(\hat{x})$$

   $$Q(\hat{x}) = I - P(\hat{x})$$

   $$V = N(\hat{x})$$

   $$V^{\perp} = T(\hat{x})$$

   then P, Q, V and $V^{\perp}$ have all the properties described

   in Exercises 1 and 2 , Chapter 7.

3. Suppose that the rocket described in the example of the
   present chapter consisted of two stages of length x and
   y, the "y" stage including the payload. Using only the

single constraint $x + y = 10$, what values of x and y

give maximum velocity increment? Solve the problem in

two ways: (i) work it as a problem in two variables and

use Theorem 9. 3 to enable you to find the optimal values;

(ii) Perform the substitution $y = 10 - x$ and work the

problem as an unconstrained maximization problem in the

single variable x.

4.   Find the rectangle of largest area whose vertices lie on

the curve $x^2 + 3y^2 = 9$. There are very easy ways and

rather hard ways to do this problem. We suggest that

you maximize xy subject to the above constraint and

use Theorem 9. 3.

5.   (Preferably a computer assignment.) Use the gradient

projection method to find the two largest eigenvalues

λ_1 and λ_2 of the symmetric matrix

$$A = \begin{pmatrix} 4 & 3 & 2 & 1 \\ 3 & 2 & 1 & 4 \\ 2 & 1 & 4 & 3 \\ 1 & 4 & 3 & 2 \end{pmatrix}.$$

To find λ_1 we maximize $x'Ax$ subject to the single constraint $x'x = 1$. To find λ_2 we maximize $x'Ax$ subject to the two constraints $x'x = 1$, $x_1'x = 0$, where $x_1 \in R^4$ is the eigenvector corresponding to the eigenvalue λ_1 which has already been found. See Chapter 7.

6. (Computer assignment). Let x_k, $k = 0, \ldots, 5$ and u_k, $k = 1, \ldots, 5$ be real numbers which satisfy

$$x_0 = 10$$

$$x_{k+1} = x_k + u_{k+1}, \quad k = 0, \ldots, 4.$$

Subject to these constraints, minimize the objective

$$5(x_5)^2 + \sum_{k=1}^{5} (u_k)^2.$$

7. Consider the optimization problem

minimize $f(x)$, $x \in S = \{x \in R^n \mid g_i(x) = 0, \ i = 1, 2, \ldots, m\}$.

Show that if f is convex and the g_i are linear, i. e.,

$g_i(x) = a_i'x + b_i$, $i = 1, 2, \ldots, m$, then the fact that the

Lagrange multiplier condition is satisfied at a point

$x^* \in S$ guarantees that x^* is a global minimum for f

in S. If the $g_i(x)$ are not linear show by example that

this need not be true even if f is strictly convex.

CHAPTER 10.   LINEAR INEQUALITY CONSTRAINTS

In this chapter we give a very brief introduction to the type of problem usually discussed under the heading of Mathematical Programming.   Our treatment barely skims the surface because this is a rather vast area with a very large body of theoretical literature and correspondingly many computer techniques.

Let $g_i : R^n \to R^1$ be continuously differentiable, $i = 1, 2, \ldots, m$.   We consider a set $S \subseteq R^n$ described by <u>inequalities</u>:

$$S = \{x \in R^n \mid g_i(x) \leq 0, \quad i = 1, 2, \ldots, m\}.$$

Given a function $f : R^n \to R^1$, also continuously differentiable, we ask for a point $x^* \in S$ such that

341

$$f(x^*) \leq f(x), \qquad x \in S,$$

that is, we seek to minimize $f(x)$ subject to the inequality

constraints $g_i(x) \leq 0$, $i = 1, 2, \ldots, m$.

As an example we consider

$$S = \{ \begin{pmatrix} x \\ y \end{pmatrix} \in R^2 \,|\, y^2 + 1 - x \leq 0, \; x^2 + y^2 - 4 \leq 0, \; y - \tfrac{1}{2}x + \tfrac{1}{2} \leq 0 \}$$

$$f(x, y) = x^2 + (y + 1)^2.$$

The region S is shaded in Figure 20. Minimizing f in S

amounts to finding that point $\begin{pmatrix} x^* \\ y^* \end{pmatrix}$ in S which lies closest

to the point $\begin{pmatrix} 0 \\ -1 \end{pmatrix}$. The solution is clearly $x^* = 1$, $y^* = 0$.

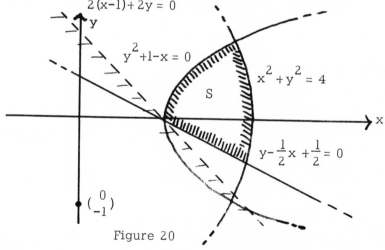

Figure 20

Just as in the case of equality constraints discussed in

Chapter 9 we note that the gradient of f does not vanish at

$\binom{x^*}{y^*}$. Observe in this case however that the open half-plane

$$\frac{\partial f}{\partial x}(x^*, y^*)(x - x^*) + \frac{\partial f}{\partial y}(x^*, y^*)(y - y^*) < 0,$$

i. e.,

$$2(x - 1) + 2y < 0,$$

has $x^* = 1$, $y^* = 0$ as a boundary point but does not itself

meet S. Observations of this nature lead to the theorems

and techniques of the present chapter.

To avoid a number of complications we will consider

only the case of <u>linear</u> inequalities in this text. This means

that the functions $g_i(x)$ are <u>linear</u>, i. e., there are vectors

$a_i \in R^n$, $i = 1, 2, \ldots, m$, and real constants b_i, $i = 1, 2, \ldots, m$,

such that

$$g_i(x) = a_i'x + b_i, \qquad i = 1, 2, \ldots, m.$$

However, we make no such restriction on the objective func-

tion f.

Linear constraints lead to sets

$$S = \{x \in R^n \mid a_i'x + b_i \leq 0, \qquad i = 1, 2, \ldots, m\}$$

which are convex <u>polyhedra</u>, i. e. , convex sets whose bound-

aries are formed by planar sections. The set

$$S = \{(\begin{smallmatrix}x\\y\end{smallmatrix}) \in R^2 \,|\, x-y-3 \le 0,\ 3x+y-13 \le 0,$$

$$3x+2y-14 \le 0,\ x+4y-20 \le 0,\ -x+4y-14 \le 0,$$

$$-3x+y-9 \le 0,\ -x-2y-3 \le 0\}$$

for which $b_1 = -3$, $b_2 = -13$, $b_3 = -14$, $b_4 = -20$, $b_5 = -14$,

$b_6 = -9$, $b_7 = -3$, and

$$a_1 = (\begin{smallmatrix}1\\-1\end{smallmatrix}),\ a_2 = (\begin{smallmatrix}3\\1\end{smallmatrix}),\ a_3 = (\begin{smallmatrix}3\\2\end{smallmatrix}),\ a_4 = (\begin{smallmatrix}1\\4\end{smallmatrix}),\ a_5 = (\begin{smallmatrix}-1\\4\end{smallmatrix}),$$

$$a_6 = (\begin{smallmatrix}-3\\1\end{smallmatrix}),\ a_7 = (\begin{smallmatrix}-1\\-2\end{smallmatrix})$$

is shown in Figure 21.

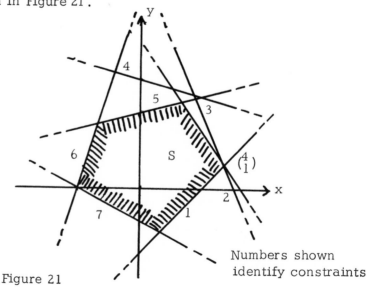

Figure 21

Numbers shown
identify constraints

It should be noted that not all of the seven inequalities which

we have listed are actually necessary to describe the set S.

The set S above remains unchanged if we delete the second

and fourth inequalities. In most practical problems the num-

ber of inequalities used to describe S is large and it would

be a very difficult task to sort out those which are superfluous.

Fortunately we are concerned with the set S one point at a

time, which makes the situation somewhat simpler.

There are two ways in which a constraint $a'x + b \leq 0$

can be satisfied at a point $\hat{x} \in S$. We can have $a'\hat{x} + b = 0$

or we can have $a'\hat{x} + b < 0$. In the first instance we say

that the constraint $a'x + b \leq 0$ is <u>active</u> at the point $\hat{x}$ while

in the second case we say that it is <u>inactive</u> at $\hat{x}$. If in the

example given above we take the point $\binom{4}{1}$ then the first,

second and third constraints are active at $\binom{4}{1}$ while the re-

maining constraints are inactive there. On the other hand,

the point $\binom{0}{0}$ lies in S but none of the constraints are

active there. Certain constraints, such as the fourth in the

above example, may not be active at any point of S.

Let us take S to be a subset of R^n described by

linear constraints $a_i'x + b_i \leq 0$, $i = 1, 2, \ldots, m$ and let $\hat{x}$

be an arbitrary point in S. We let $I(\hat{x})$ be the subset of the

integers $1, 2, \ldots, m$ which indicates the active constraints at

$\hat{x}$, that is, the constraint $a_i'x + b \le 0$ is active at $\hat{x} \in S$ if

and only if $i \in I(\hat{x})$. In agreement with the notation used in

the previous chapter we will let $\Gamma(\hat{x})$ be the matrix of dimen-

sion $m(\hat{x})$ by n, $m(\hat{x})$ being the number of constraints

active at $\hat{x}$, whose rows are a_i', $i \in I(\hat{x})$. Thus for the point

$\binom{4}{1}$ in the example above we have

$$\Gamma\binom{4}{1} = \begin{pmatrix} 1 & -1 \\ 3 & 1 \\ 3 & 2 \end{pmatrix}$$

On the other hand

$$\Gamma\binom{-1}{-1} = (-1, -2)$$

since only the seventh constraint is active there. Finally

$\Gamma\binom{0}{0}$ is void, because no constraints are active at the origin

in our example.

It will be necessary to further distinguish between

types of points in S in the following way. We will say that

$\hat{x}$ is a regular point of S if the rows of $\Gamma(\hat{x})$ are linearly

independent or if $\Gamma(\hat{x})$ is void. If the rows of $\Gamma(\hat{x})$ are

linearly dependent we will say that $\hat{x}$ is a point of degeneracy.

In the example above $\binom{-1}{-1}$ and $\binom{2}{4}$ are both regular points while $\binom{4}{1}$ is a point of degeneracy. Regular points are by far the most common. The existence of any point of degeneracy is an exceptional occurrence but cannot be ruled out. The point $\binom{4}{1}$ is the only point of degeneracy in the above example.

When working with a digital computer or any other system with limited precision the question of whether a given constraint is active at a point or not and the question of whether a point is regular or degenerate may not be entirely clear cut. Generally speaking it depends upon what accuracy we are striving for. Thus, in the above example we might consider the third and fifth constraints to both be active at the point $\binom{1.99}{3.99}$ if we allow a tolerance $\pm.01$ whereas if we insist upon a tolerance $\pm.0001$ then no constraints are considered active there. Best practice, after setting a tolerance $\pm \epsilon$, is to compute $(a_i' \hat{x} + b_i)/\|a_i'\|$ (any norm) and consider the constraint active if and only if the absolute value of this quantity is $\leq \epsilon$. The question of degeneracy is slightly more complicated. If, using the criteria just described, we find that more than n constraints are active at $\hat{x}$

then it is clear that $\hat{x}$ is a point of degeneracy. If the number of constraints is $\leq n$ one could take $\hat{x}$ to be a point of degeneracy if

$$|\det G(\hat{x})| = |\det(\Gamma(\hat{x})\Gamma(\hat{x})')| \leq \epsilon$$

and a regular point otherwise. Strictly speaking it would be best again to use the vectors $\dfrac{a_i'}{\|a_i\|}$ as rows of $\Gamma(\hat{x})$. In fact, whenever convenient, one should restate the description of S in terms of inequalities $\tilde{a}_i'x + \tilde{b}_i = (\dfrac{a_i}{\|a_i\|})'x + \dfrac{b_i}{\|a_i\|} \leq 0$ at the outset and use the vectors $\tilde{a}_i$ and the numbers $\tilde{b}_i$ thereafter in all computation. This is sometimes inconvenient in hand computation as we shall see.

We will again in this chapter refer to the normal and tangent spaces $N(\hat{x})$ and $T(\hat{x})$, defined as follows:

$$N(\hat{x}) = \{y \in R^n \mid y = \sum_{i \in I(x)} \alpha^i a_i, \quad \alpha^i \text{ real}\}$$

and

$$T(\hat{x}) = \{z \in R^n \mid y'z = 0, \quad y \in N(\hat{x})\}.$$

The space $N(\hat{x})$ is generated by the vectors a_i, $i \in I(\hat{x})$, which form a <u>basis</u> for $N(\hat{x})$ if $\hat{x}$ is a regular point. The situation is then similar to that of Chapter 9 where we assumed the algebraic constraint qualification. As we have indicated, this is the typical situation as points of degeneracy are exceptional. Many of the results presented below will be proved only for the regular case but, unless we explicitly state to the contrary, they also hold in the case of degeneracy.

DEFINITION 10.1. Let $\hat{x}$ be a regular point of the set S so that the vectors a_i, $i \in I(\hat{x})$ are linearly independent and thus form a basis for $N(\hat{x})$. A second set of vectors p_i, $i \in I(\hat{x})$ in $N(\hat{x})$ is said to be the <u>dual basis</u> for $N(\hat{x})$ corresponding to the basis a_i, $i \in I(\hat{x})$ if for $i \in I(\hat{x})$, $j \in I(\hat{x})$

$$a_i' p_j = \begin{cases} 1 & \text{if} \quad i = j \\ 0 & \text{if} \quad i \neq j \ . \end{cases}$$

We will call this the <u>biorthogonality property</u> of the dual basis.

To validate this definition we should establish that the vectors p_j do indeed form a basis for $N(\hat{x})$. In the process we will develop a method for finding the p_j. If $p_j \in N(\hat{x})$

then there are real coefficients β_j^ℓ such that

$$p_j = \sum_{\ell \in I(\hat{x})} \beta_j^\ell a_\ell, \qquad j \in I(\hat{x}).$$

Then

$$a_i' p_j = \sum_{i \in I(\hat{x})} \beta_j^\ell a_i' a_\ell$$

and the biorthogonality property gives

$$\sum_{i \in I(\hat{x})} a_i' a_\ell \beta_j^\ell = \begin{cases} 1 & \text{if} \quad i = j \\ 0 & \text{if} \quad i \neq j \end{cases}.$$

We let $G(\hat{x}) = \Gamma(\hat{x})\Gamma(\hat{x})'$ be the $m(\hat{x})$ by $m(\hat{x})$ matrix with entries $a_i' a_\ell$, $i \in I(\hat{x})$, $\ell \in I(\hat{x})$ and we let $B(\hat{x})$ be the $m(\hat{x})$ by $m(\hat{x})$ matrix with entries β_j^ℓ, $\ell \in I(\hat{x})$, $j \in I(\hat{x})$. Letting I be the $m(\hat{x})$ by $m(\hat{x})$ identity matrix we have, using the definition of matrix multiplication,

$$G(\hat{x})B(\hat{x}) = I$$

so that $B(\hat{x}) = G(\hat{x})^{-1}$, which exists if x is a regular point, as explained in Chapter 9. The formula $p_j = \sum_{\ell \in I(\hat{x})} \beta_j^\ell a_\ell$ then becomes,

$$D(\hat{x}) = \Gamma(\hat{x})'B(\hat{x}) = \Gamma(\hat{x})'G(\hat{x})^{-1}$$

where $D(\hat{x})$ is the n by $m(\hat{x})$ matrix whose columns are the vectors p_j, $j \in I(\hat{x})$. Then since $G(\hat{x})$ and $G(\hat{x})^{-1}$ are both non-singular the number of linearly independent columns of $D(\hat{x})$ (i.e., the rank of $D(\hat{x})$) is the same as the number of linearly independent rows of $\Gamma(\hat{x})$, i.e., $m(\hat{x})$, since we are in the regular case. Therefore the vectors p_j, $j \in I(\hat{x})$, constitute $m(\hat{x})$ linearly independent vectors in the $m(\hat{x})$ dimensional space $N(\hat{x})$ and, as shown in Chapter 7, constitute a basis for $N(\hat{x})$.

Exercise 8 at the end of this chapter will help the student develop a feeling for the geometric relationship of the vectors p_j of the dual basis to the vectors a_i of the original basis of $N(\hat{x})$.

THEOREM 10.1 (Kuhn Tucker Necessary Condition). Let $f : R^n \rightarrow R^1$ be continuously differentiable and let

$$S = \{x \in R^n \,|\, a_i'x + b_i \le 0, \quad i = 1, 2, \ldots, m\}.$$

If $x^* \in S$ is a local minimum for f in S, i.e., if for some

$\rho > 0$ we have

$$f(x^*) \leq f(x), \quad x \in S, \quad \|x - x^*\| < \rho$$

then there are non-negative real numbers μ^i, $i \in I(x^*)$, called

Lagrange multipliers, such that

$$\frac{\partial f}{\partial x}(x^*) + \sum_{i \in I(x^*)} \mu^i a_i' = 0.$$

<u>Proof.</u> We give the proof only for the case where x^* is a

regular point but the result is also true in the case of degen-

eracy.

Since $R^n = N(x^*) \oplus T(x^*)$ we have

$$-\frac{\partial f}{\partial x}(x^*)' = y + z, \quad y \in N(x^*), \quad z \in T(x^*).$$

If $y \in N(x^*)$ we can expand it in terms of the basis a_i,

$i \in I(x^*)$:

$$y = \sum_{i \in I(x^*)} \mu^i a_i, \quad \mu^i \text{ real}, \quad i \in I(x^*).$$

Now let $I(x^*) = I^+(x^*) \cup I^-(x^*)$ where $I^-(x^*)$ consists of

those $i \in I(x^*)$ for which μ^i is negative and $I^+(x^*)$ includes

those $i \in I(x^*)$ for which μ^i is non-negative. Then we put

$$r = \sum_{j \in I^-(x^*)} \mu^j p_j + z \, .$$

We claim that for small $\lambda > 0$ the point

$$x = x^* + \lambda r$$

lies in S. For if $i \in I(x^*)$ then $a_i' x^* + b_i = 0$ and we have, using the biorthogonality property of the p_j together with the fact that the vectors $a_i \in N(x^*)$ are orthogonal to $z \in T(x^*)$,

$$a_i' x + b_i = a_i'(x^* + \lambda (\sum_{j \in I^-(x^*)} \mu^j p_j + z)) + b_i$$

$$= a_i' x^* + b_i + \begin{cases} \lambda \mu^i & \text{if} \quad i \in I^-(x^*) \\ 0 & \text{if} \quad i \in I^+(x^*) \end{cases} \leq 0$$

in either case. On the other hand if the constraint $a_i' x + b_i \leq 0$ is not active at x^* then $a_i' x^* + b_i = -\delta_i$, $\delta_i > 0$. Then

$$a_i' x + b_i = a_i'(x^* + \lambda r) + b_i = -\delta_i + \lambda a_i' r$$

$$\leq -\delta_i + \lambda \| a_i \|_e \| r \|_e$$

and thus $a_i' x + b_i \leq 0$ providing

$$\lambda \leq \frac{\delta_i}{\| a_i \|_e \| r \|_e} \, .$$

Since there are only finitely many of these inequalities to be satisfied there is a number $\rho_0 > 0$ such that if $0 < \lambda < \rho_0$ all of these inequalities hold. Thus we conclude $x = x^* + \lambda r \in S$ if $\lambda > 0$ is sufficiently small.

Now

$$-\frac{\partial f}{\partial x}(x^*)r = (\sum_{i \in I(x^*)} \mu^i a_i + z)(\sum_{j \in I^-(x^*)} \mu^j p_j + z)$$

$$= \sum_{j \in I^-(x^*)} (\mu^j)^2 + \|z\|_e^2 = d \geq 0,$$

where we have again used the biorthogonality property and the fact that $a_i'z = 0$, $i \in I(x^*)$. Expanding f about x^* we have

$$f(x) = f(x^* + \lambda r) = f(x^*) + \lambda \frac{\partial f}{\partial x}(x^*)r + \mathcal{O}(\lambda)$$

$$= f(x^*) - \lambda d + \mathcal{O}(\lambda).$$

Using an argument by now familiar to us we see that if we are to have $f(x^*) \leq f(x)$ for small $\lambda > 0$ then we must have $d \leq 0$. But $d \geq 0$ as shown above and thus

$$d = \sum_{j \in I^-(x^*)} (\mu^j)^2 + \|z\|_e^2 = 0.$$

The only way this can be true is $z = 0$ and $I^-(x^*)$ void.

Thus all μ^i are non-negative and

$$-\frac{\partial f}{\partial x}(x^*) = \sum_{i \in I(x^*) = I^+(x^*)} \mu^i a_i,$$

which completes the proof of the theorem.

The complexity of the necessity argument above should be compared with the ease of the sufficiency theorem below.

THEOREM 10.2 (Kuhn-Tucker Sufficiency Condition). Let the Kuhn-Tucker condition

$$\frac{\partial f}{\partial x}(x^*) + \sum_{i \in I(x^*)} \mu^i a_i', \quad \mu \geq 0, \quad i \in I(x^*)$$

hold at $x^* \in S$, where $f : R^n \to R^1$ is continuously differentiable and convex in S. Then

$$f(x^*) \leq f(x), \quad x \in S.$$

Proof. Since f is convex in S, Theorem 6.5 gives

$$f(x) \geq f(x^*) + \frac{\partial f}{\partial x}(x^*)(x - x^*)$$

$$= f(x^*) - \sum_{i \in I(x^*)} \mu^i a_i'(x - x^*).$$

Now $a_i' x^* = - b_i$, $i \in I(x^*)$, and $a_i' x \leq - b_i$ so we have $a_i'(x - x^*) \leq 0$. The non-negativity of the μ^i then gives $- \mu^i a_i'(x - x^*) \geq 0$ and thus

$$f(x) \geq f(x^*) - \sum_{i \in I(x^*)} \mu^i a_i'(x - x^*) \geq f(x^*)$$

and the proof is complete.

It should be noted that the Kuhn Tucker condition reduces to the condition $\dfrac{\partial f}{\partial x}(x^*) = 0$ if none of the constraints are active at the point x^*. This is in agreement with Theorem 5.2 of Chapter 5.

We shall consider two simple examples, in both cases taking S to be the set shown in Figure   , i.e.,

$$S = \{ \binom{x}{y} \in R^2 \,|\, x - y - 3 \leq 0, \; 3x + y - 13 \leq 0,$$

$$3x + 2y - 14 \leq 0, \; x + 4y - 20 \leq 0, \; -x + 4y - 14 \leq 0,$$

$$-3x + y - 9 \leq 0, \; -x - 2y - 3 \leq 0 \}.$$

For the first example we let

$$f(x, y) = x + y.$$

We claim that the point $\binom{2}{4} \in S$ is a global maximum for f

in S. The third and fifth constraints, $3x + 2y - 14 \leq 0$ and

$-x + 4y - 14 \leq 0$, with gradients $(3, 2)$ and $(-1, 4)$ respec-

tively, are active at this point. The gradient of $-f$ is $(-1, -1)$.

The equation

$$(-1, -1) + \mu^3(3, 2) + \mu^5(-1, 4) = 0$$

has exactly one solution: $\mu^3 = \dfrac{5}{14}$ and $\mu^5 = \dfrac{1}{14}$. Thus,

since $-f$ is convex, the conditions of Theorem 10. 2 are

satisfied at $\binom{2}{4}$ and we have there a global minimum for $-f$,

i. e. , a global maximum for f , relative to other points in S.

For our second example we take

$$f(x, y) = x^2 + y^2$$

and again seek a maximum. In this case $-f$ is not convex

and Theorem 10. 2 cannot be applied. In fact, for this problem

there are no less than 5 local maxima for f in S: each of

the five vertices of S is a local maximum. The reader may

verify that the Kuhn Tucker conditions are satisfied at each

of these points, in agreement with the necessity conditions

of Theorem 10. 1. Let us consider the situation at $\binom{4}{1}$, which

is a point of degeneracy. The gradient of $-f$ here is $(-8, 2)$.

There are three active constraints at $\binom{4}{1}$, the first, second

and third, with gradients $(3,-3)$, $(3,1)$ and $(3,2)$, respec-

tively. The equation

$$(-8,-2) + \mu^1(3,-3) + \mu^2(3,1) + \mu^3(3,2) = 0$$

has infinitely many solutions. In fact we can choose any one

of μ^1, μ^2, μ^3 arbitrarily and then solve for the other two. If

we take $\mu^2 = 0$ then

$$(-8,-2) + \mu^1(3,-3) + \mu^3(3,2) = 0$$

yields $\mu^1 = \frac{2}{3}$, $\mu^3 = 2$. These values provide verification

for Theorem 10.1. If we put $\mu^2 = 1$ then

$$(-8,-2) + \mu^1(3,-3) + (3,1) + \mu^3(3,2) = 0$$

yields $\mu^1 = \frac{7}{15}$, $\mu = \frac{6}{5}$, again in agreement with Theorem 10.1.

We see therefore that in the case of degeneracy the μ^i need

not be unique. With $\mu^1 = \frac{13}{15}$, $\mu^2 = -1$, $\mu^3 = \frac{14}{5}$ we have a

solution of the equation with μ^2 negative. This is unimpor-

tant since Theorem 10.1 merely asserts there is some solution

with all μ^i non-negative, it does not guarantee that all solu-

tions will have this property.

The Kuhn Tucker condition

$$\frac{\partial f}{\partial x}(x^*) + \sum_{i \in I(x^*)} \mu^i a_i^i = 0, \quad \mu^i \geq 0, \quad i \in I(x^*)$$

guarantees only that the open half-space $\frac{\partial f}{\partial x}(x^*)(x - x^*) < 0$

does not intersect S. (The student is asked to prove this in

Exercise 2). This is a necessary condition. If f is not con-

vex we cannot be sure that the satisfaction of the Kuhn-Tucker

conditions at x^* means that x^* is even a local minimum.

Consider for example the problem

$$\text{minimize } f(x, y) = -x + y^3$$

$$\binom{x}{y} \in S = \{\binom{x}{y} \in R^2 \mid x \leq 0\}.$$

At the origin the gradient of f is $(-1, 0)$ while the gradient

of the single constraint is $(1, 0)$ and with $\mu^1 = 1$ we satisfy

$$(-1, 0) + \mu^1(1, 0) = 0.$$

However the origin is not a local minimum for f in S.

We will now describe a method, <u>the gradient projection

technique for inequality constraints</u>, which will enable us to

approximate solutions of optimization problems of the type

discussed in Theorems 10.1 and 10.2. The procedure is quite similar to those discussed in Chapters 8 and 9. A sequence of points in S, $\{x_k\}$, is generated by recursion relations

$$x_{k+1} = x_k + \lambda_k r_k,$$

where r_k is a <u>feasible direction</u> at x_k and λ_k is the <u>step parameter</u> which determines how far one should move in the feasible direction. Unlike the situation in Chapters 8 and 9 we do not take λ_k to be a fixed constant in this chapter.

The discussion of the method naturally divides into two parts: the determination of a <u>feasible direction</u> r_k at a point x_k and the determination of the step parameter λ_k. We will treat these in the order indicated.

DEFINITION 10.2. Let $\hat{x} \in S$ and let the active constraints at $\hat{x}$ be $a_i' x + b_i \leq 0$, $i \in I(\hat{x})$. A vector $\hat{r} \in R^n$ gives a <u>feasible direction</u> at $\hat{x}$ if

$$a_i' \hat{r} \leq 0, \qquad i \in I(\hat{x}),$$

$$\frac{\partial f}{\partial x}(\hat{x})\hat{r} < 0.$$

Equivalently, as shown in Theorem 10.1, $\hat{r}$ is a feasible dir-

ection at $\hat{x}$ if $\hat{x} + \lambda \hat{r} \in S$ and $f(\hat{x} + \lambda \hat{r}) < f(\hat{x})$ for small posi-

tive values of λ .

Now, given a point $\hat{x} \in S$, how does one determine a

feasible direction r ? The first step would be to try $\hat{r} =$

$- \frac{\partial f}{\partial x} (\hat{x})'$ as in the ordinary gradient method. This certainly

is the thing to do if no constraints are active at $\hat{x}$. Even if

some constraints are active there it may happen that

$a_i'(- \frac{\partial f}{\partial x} (\hat{x})') \leq 0$, $i \in I(\hat{x})$, so that the direction $- \frac{\partial f}{\partial x} (\hat{x})'$ re-

mains feasible.

If this preliminary step fails we turn to the procedure

already outlined in Theorem 10.1. We will assume for the time

being that $\hat{x}$ is a regular point, postponing the discussion of

degeneracy until later.

The first step is to calculate the dual basis p_j ,

$j \in I(\hat{x})$, with the property

$$a_i' p_j = \begin{cases} 1 & \text{if} \quad i = j \\ 0 & \text{if} \quad i \neq j \end{cases} \quad i, j \in I(\hat{x}).$$

As we have already shown, if one takes $D(\hat{x})$ to be the matrix

whose columns are the vectors p_j , then

$$D(\hat{x}) = \Gamma(\hat{x})'G(\hat{x})^{-1}$$

where $\Gamma(\hat{x})$ is the matrix whose rows are the vectors a_i, $i \in I(\hat{x})$, and $G(\hat{x}) = \Gamma(\hat{x})\Gamma(\hat{x})'$. If $\hat{x}$ is a regular point then $G(\hat{x})^{-1}$ exists and the p_j are found without difficulty. One should observe, however, that if there are n active constraints at $\hat{x} \in S \subseteq R^n$ then $\Gamma(\hat{x})$ is itself non-singular and the above formula reduces to

$$D(\hat{x}) = \Gamma(\hat{x})^{-1}.$$

Once the vectors p_j, $j \in I(\hat{x})$ have been found the rest is easy. We have, since $R^n = N(\hat{x}) \oplus T(\hat{x})$

$$-\frac{\partial f}{\partial x}(\hat{x})' = y + z = \sum_{i \in I(\hat{x})} \mu^i a_i + z$$

for some real numbers μ^i, $i \in I(\hat{x})$. The numbers μ^i are readily calculated if we use the biorthogonal property of the p_j relative to the a_i together with the fact that $z'p_j = 0$, $j \in I(\hat{x})$ (which follows from the fact that $z \in T(\hat{x})$ while $p_j \in N(\hat{x})$). Thus

$$-\frac{\partial f}{\partial x}(\hat{x})p_j = \sum_{i \in I(\hat{x})} \mu^i a_i'p_j + z'p_j = \mu^j, \qquad j \in I(\hat{x}).$$

Then

$$\hat{y} = \sum_{i \epsilon I(\hat{x})} (- \frac{\partial f}{\partial x} (\hat{x}) p_i) a_i$$

and

$$\hat{z} = - \frac{\partial f}{\partial x} (\hat{x})' - \hat{y} = - \frac{\partial f}{\partial x} (\hat{x})' + \sum_{i \epsilon I(\hat{x})} (\frac{\partial f}{\partial x} (\hat{x}) p_i) a_i .$$

If $\hat{z} = 0$ and all $\mu^i \geq 0$ then the Kuhn-Tucker conditions are satisfied at $\hat{x}$ and we would take this as an indication that $\hat{x}$ might well be the desired minimum of f in S, in fact if Theorem 10.2 applies we can be sure of this. One can easily verify in this case that no feasible direction $\hat{r}$ exists.

If $\hat{z} \neq 0$ and/or some $\mu^i < 0$ we put

$$\hat{r} = \sum_{\substack{i \epsilon I(\hat{x}) \\ \mu^i < 0}} \mu^i p_i + z$$

$$= \sum_{i \epsilon I(\hat{x})} (- \frac{\partial f}{\partial x} (\hat{x}) p_i) p_i - \frac{\partial f}{\partial x} (\hat{x})'$$

$$- \frac{\partial f}{\partial x} (\hat{x}) p_i < 0$$

$$+ \sum_{i \epsilon I(\hat{x})} (\frac{\partial f}{\partial x} (\hat{x}) p_i) a_i .$$

As shown in the proof of Theorem 10.1, $\hat{r}$ as thus computed is a feasible direction. We are assured that for small values of $\lambda > 0$ we will have $a_i'\hat{r} \leq 0$, $i \in I(\hat{x})$, and

$$\frac{\partial f}{\partial x}(\hat{x})\hat{r} = -\left(\sum_{\substack{i \in I(\hat{x}) \\ \mu^i < 0}} (\mu^i)^2 + \|\hat{z}\|_e^2 \right) < 0.$$

We pass now to a new point $\hat{x} + \lambda\hat{r}$, where $\lambda > 0$ and $\hat{r}$ is the feasible direction chosen as described above. The question is, what value should we use for λ? First of all the available values of λ will in most cases be limited by the inequality constraints. If $\hat{x} + \lambda\hat{r}$ is to belong to S we must have

$$a_i'(\hat{x} + \lambda\hat{r}) + b_i \leq 0, \qquad i = 1, 2, \ldots, m.$$

Since $a_i'\hat{x} + b_i = 0$ and $a_i'\hat{r} \leq 0$ for $i \in I(\hat{x})$ we do not have to worry about those constraints which are active at $\hat{x}$. If we let $J(\hat{x})$ be those indices in the set $\{1, 2, \ldots, m\}$ which do not lie in $I(\hat{x})$, i.e., $J(\hat{x}) = \{1, 2, \ldots, m\} - I(\hat{x})$ then we need

$$a_i'(\hat{x} + \lambda\hat{r}) + b_i \leq 0, \qquad i \in J(\hat{x}).$$

If $a_i'\hat{r} \leq 0$ this provides no new restriction on λ. If $a_i'\hat{r} > 0$ we must have

$$\lambda \leq -\frac{(a_i'\hat{x} + b_i)}{a_i'\hat{r}} \ .$$

We therefore let

$$\tilde{\lambda} = \min_{\substack{i \in J(\hat{x}) \\ a_i'\hat{r} > 0}} \frac{-(a_i'\hat{x} + b_i)}{a_i'\hat{r}} \ ,$$

provided the set $\{i \mid i \in J(\hat{x}), a_i'\hat{r} > 0\}$ is not empty. Note that $\tilde{\lambda}$ will be positive since $a_i'\hat{x} + b_i > 0$ when $i \in J(\hat{x})$, for $J(\hat{x})$ is the set of indices corresponding to the inactive constraints at $\hat{x}$. If the set $\{i \mid i \in J(\hat{x}), a_i'\hat{r} > 0\}$ is empty then $\tilde{\lambda}$ cannot be defined this way. (This cannot happen if S is a bounded region but is possible if S is unbounded.) In this case one should either set $\tilde{\lambda} = +\infty$ or else use some nominal value, e.g., $\tilde{\lambda} = 10$, 100, etc., depending on the scale of the problem at hand.

Having found $\tilde{\lambda} > 0$ we know that

$$\hat{x} + \lambda\hat{r} \in S, \quad 0 \leq \lambda \leq \tilde{\lambda} \ .$$

We now attempt to choose $\lambda > 0$ in such a way that $\hat{x} + \lambda r$

minimizes f relative to points $x + \lambda r$, $0 \leq \lambda \leq \tilde{\lambda}$. This is

actually a one-dimensional minimization problem, namely

$$\text{minimize } \hat{f}(\lambda) = f(\hat{x} + \lambda \hat{r}), \qquad 0 \leq \lambda \leq \tilde{\lambda}.$$

Depending on circumstances, there are a number of ways in

which we might solve this auxiliary one-dimensional problem.

The case where the objective function f is linear is quite

special and will be discussed separately.

Let us consider the special case of a linear-quadratic

objective

$$f(x) = x'Px + q'x,$$

where P is an $n \times n$ symmetric positive semi-definite ma-

trix and q is an n-vector. We compute

$$f(\hat{x} + \lambda \hat{r}) = (\hat{r}'P\hat{r})\lambda^2 + (2\hat{x}'P\hat{r} + q'\hat{r})\lambda + \hat{x}'P\hat{x} + q'\hat{x}.$$

Therefore

$$\frac{\partial f}{\partial \lambda}(\hat{x} + \lambda \hat{r}) = 2\hat{r}'P\hat{r}\lambda + 2\hat{x}'P\hat{r} + q'\hat{r}.$$

If $\hat{r}'P\hat{r} = 0$ then $\frac{\partial f}{\partial \lambda}(\hat{x} + \lambda \hat{r})$ is identically equal to the negative constant $2\hat{x}P\hat{r} + q'\hat{r}$ (negative because $\frac{\partial f}{\partial x}(\hat{x})\hat{r} < 0$)

and the minimum occurs at the endpoint $\lambda = \hat{\lambda} = \tilde{\lambda}$ if $\tilde{\lambda}$ is

finite or, if $\tilde{\lambda} = \infty$, there is no minimum and the original min-

imization problem has no solution. If $\hat{r}'P\hat{r} > 0$, then

$\frac{\partial f}{\partial \lambda}(\hat{x} + \lambda\hat{r}) = 0$ if and only if

$$\lambda = \frac{-2\hat{x}'P\hat{r} - q'\hat{r}}{2\hat{r}'P\hat{r}}$$

and we put

$$\hat{\lambda} = \min\left\{\tilde{\lambda}, \frac{-2\hat{x}'P\hat{r} - q'\hat{r}}{2\hat{r}'P\hat{r}}\right\} .$$

If we have reason to believe that $\hat{f}(\lambda) = f(\hat{x} + \lambda\hat{r})$ is

$(-)$ unimodal in $[0, \tilde{\lambda}]$ we might use the one-dimensional

search techniques of Chapter 3 to locate $\hat{\lambda}$ in the interval

$[0, \tilde{\lambda}]$.

For general objective functions it is not easy to give

a foolproof formula for finding $\hat{\lambda}$. We may, in fact, have to

be content with a local minimum of $\hat{f}(\lambda) = f(\hat{x} + \lambda\hat{r})$. The

gradient method

$$\lambda_{k+1} = \lambda_k - \gamma\hat{f}'(\lambda_k) = \lambda_k - \gamma\frac{\partial f}{\partial x}(\hat{x} + \lambda_k\hat{r})\hat{r} ,$$

with a suitable positive γ and starting at $\lambda_0 = 0$ could be used to find the first value of $\lambda > 0$ where $\hat{f}'(\lambda) = 0$, provided this value does not exceed $\tilde{\lambda}$. We might then use this value for $\hat{\lambda}$. We would definitely have $f(\hat{x} + \lambda\hat{r}) < f(\hat{x})$ but we could not be certain of a minimum.

Generally speaking, it is not worthwhile to put too much effort into locating $\hat{\lambda}$ with great accuracy. Just how much to strive for is a matter of judgment. One "quick and dirty" technique is to put

$$\lambda_0 = \tilde{\lambda}$$

$$\lambda_{k+1} = \frac{\lambda_k}{2}, \qquad k \geq 0$$

and set $\hat{\lambda} = \lambda_k$ the first time the inequalities

$$f(\hat{x} + \lambda_k\hat{r}) < f(\hat{x} + \lambda_{k+1}\hat{r}) < f(\hat{x})$$

are satisfied, as they eventually must be for sufficiently large k.

We are ready now to present our first examples of the use of the gradient projection technique for problems with inequality constraints. We do so with two very simple problems, both involving the set

$$S = \{ \begin{pmatrix} x \\ y \\ z \end{pmatrix} \in R^3 \mid z - 2 \leq 0, \; -z-2 \leq 0,$$

$$2x + 2y + z - 10 \leq 0, \quad -2x - 2y - z - 10 \leq 0$$

$$2x - y - z - 8 \leq 0, \quad -2x + y + z - 8 \leq 0 \}.$$

Thus

$$a_1 = \begin{pmatrix} 0 \\ 0 \\ 1 \end{pmatrix}, \quad a_2 = \begin{pmatrix} 0 \\ 0 \\ -1 \end{pmatrix}, \quad a_3 = \begin{pmatrix} 2 \\ 2 \\ 1 \end{pmatrix}, \quad a_4 = \begin{pmatrix} -2 \\ -2 \\ -1 \end{pmatrix},$$

$$a_5 = \begin{pmatrix} 2 \\ -1 \\ -1 \end{pmatrix}, \quad a_6 = \begin{pmatrix} -2 \\ 1 \\ 1 \end{pmatrix},$$

$$b_1 = -2, \; b_2 = -2, \; b_3 = -10, \; b_4 = -10, \; b_5 = -8, \; b_6 = -8.$$

For the first example we will use the objective function

$$f(x, y, z) = (x + 3)^2 + (y + 3)^2 + (z + 3)^2$$

while for the second we will use

$$h(x, y, z) = (y - 10)^2.$$

In both cases we will be able to obtain the minimum in just

two steps.   However, essentially all cases which can arise

in the use of the gradient projection technique will be covered

in these examples. We will not normalize the vectors a_i so

that $\|a_i\|_e = 1$ because it would make our hand calculations

more difficult. We will use exact fractional arithmetic so that

we do not need to be concerned with rounding errors. The set

S is shown in Figure 22 together with the paths followed by

the gradient projection technique in both cases.

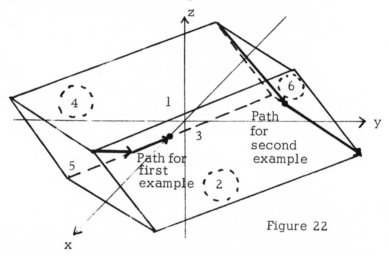

Figure 22

EXAMPLE. $f(x,y,z) = (x+3)^2 + (y+3)^2 + (z+3)^2$.

In this very simple example we find a feasible point

by treating the first third and fifth inequalities as equalities

$$z = 2$$

$$2x + 2y + z = 10$$

$$2x - y - z = 8 .$$

The solution $x_0 = \frac{14}{3}$, $y_0 = -\frac{2}{3}$, $z_0 = 2$ is readily seen to satisfy the other inequalities, i. e. , the second, fourth and sixth, as strict inequalities.

Step 1.
$$\begin{pmatrix} x_0 \\ y_0 \\ z_0 \end{pmatrix} = \begin{pmatrix} \frac{14}{3} \\ -\frac{2}{3} \\ 2 \end{pmatrix}, \quad I\begin{pmatrix} x_0 \\ y_0 \\ z_0 \end{pmatrix} = \{1, 3, 5\},$$

We compute the negative gradient at the initial point:

$$(-2(\frac{14}{3} + 3), \ -2(-\frac{2}{3} + 3), \ -2(2 + 3)) = (-\frac{46}{3}, -\frac{14}{3}, -10).$$

We begin by forming the scalar product of the negative gradient with the vectors a_1, a_3, a_5:

$$(0, 0, 1) \begin{pmatrix} -\frac{46}{3} \\ -\frac{14}{3} \\ -10 \end{pmatrix} = -10, \quad (2, 2, 1) \begin{pmatrix} -\frac{46}{3} \\ -\frac{14}{3} \\ -10 \end{pmatrix} = -50,$$

$$(2, -1, 1) \begin{pmatrix} -\frac{46}{3} \\ -\frac{14}{3} \\ -10 \end{pmatrix} = -16.$$

Since all of these are negative, the negative gradient is itself an admissible direction at the initial point.   Thus

$$r_0 = \begin{pmatrix} -\dfrac{46}{3} \\ -\dfrac{14}{3} \\ -10 \end{pmatrix}$$

We next compute $\tilde{\lambda}$ as indicated previously, using only those

constraints inactive at the initial point.   We compute:

For $i = 2$,
$$\frac{-\left[(0, 0, -1)\begin{pmatrix} \dfrac{14}{3} \\ -\dfrac{2}{3} \\ 2 \end{pmatrix} - 2 \right]}{(0, 0, -1)\begin{pmatrix} -\dfrac{46}{3} \\ -\dfrac{14}{3} \\ -10 \end{pmatrix}} = \frac{2}{5} \ ;$$

For $i = 4$,
$$\frac{-\left[(-2, -2, -1)\begin{pmatrix} \dfrac{14}{3} \\ -\dfrac{2}{3} \\ 2 \end{pmatrix} - 10 \right]}{(-2, -2, -1)\begin{pmatrix} -\dfrac{46}{3} \\ -\dfrac{14}{3} \\ -10 \end{pmatrix}} = \frac{2}{5} \ ;$$

For $i = 6$,

$$-\left[(-2, 1, 1) \begin{pmatrix} \dfrac{14}{3} \\ -\dfrac{2}{3} \\ 2 \end{pmatrix} - 8 \right] \bigg/ (-2, 1, 1) \begin{pmatrix} -\dfrac{46}{3} \\ -\dfrac{14}{3} \\ -10 \end{pmatrix} = 1 \; .$$

Thus $\tilde{\lambda} = \dfrac{2}{5}$. To minimize $f\left(\begin{pmatrix} \dfrac{14}{3} \\ -\dfrac{2}{3} \\ 2 \end{pmatrix} + \lambda \begin{pmatrix} -\dfrac{46}{3} \\ -\dfrac{14}{3} \\ -10 \end{pmatrix} \right)$ for

$0 \leq \lambda \leq \dfrac{2}{5}$ we note that f is quadratic and compute, in agreement with the method described earlier, taking $P = I$, $q' = (6, 6, 6)$

$$\frac{-2\left(\dfrac{14}{3}, -\dfrac{2}{3}, 2\right) \begin{pmatrix} 1 & 0 & 0 \\ 0 & 1 & 0 \\ 0 & 0 & 1 \end{pmatrix} \begin{pmatrix} -\dfrac{46}{3} \\ -\dfrac{14}{3} \\ -10 \end{pmatrix} - (6, 6, 6) \begin{pmatrix} -\dfrac{46}{3} \\ -\dfrac{14}{3} \\ -10 \end{pmatrix}}{2\left(-\dfrac{46}{3}, -\dfrac{14}{3}, -10\right) \begin{pmatrix} 1 & 0 & 0 \\ 0 & 1 & 0 \\ 0 & 0 & 1 \end{pmatrix} \begin{pmatrix} -\dfrac{46}{3} \\ -\dfrac{14}{3} \\ -10 \end{pmatrix}} = \frac{1}{2} \; .$$

Since $\dfrac{2}{5} < \dfrac{1}{2}$ we put $\lambda_0 = \dfrac{2}{5}$ and our new point is

$$\begin{pmatrix} x_1 \\ y_1 \\ z_1 \end{pmatrix} = \begin{pmatrix} \dfrac{14}{3} \\ -\dfrac{2}{3} \\ 2 \end{pmatrix} + \dfrac{2}{5} \begin{pmatrix} -\dfrac{46}{3} \\ -\dfrac{14}{3} \\ -10 \end{pmatrix} = \begin{pmatrix} -\dfrac{22}{15} \\ -\dfrac{38}{15} \\ -2 \end{pmatrix}.$$

Step 2.   The active constraints at this new point are readily

seen to be specified by

$$I \begin{pmatrix} -\dfrac{22}{15} \\ -\dfrac{38}{15} \\ -2 \end{pmatrix} = \{2, 4\}.$$

The negative gradient of f at the new point is

$$\left(-2\left(-\dfrac{22}{15} + 3\right),\ -2\left(-\dfrac{38}{15} + 3\right),\ -2(-2 + 3)\right) = \left(-\dfrac{46}{15},\ -\dfrac{14}{15},\ -2\right).$$

Taking the scalar product with a_2, a_4 :

$$(0, 0, -1) \begin{pmatrix} -\dfrac{46}{15} \\ -\dfrac{14}{15} \\ -2 \end{pmatrix} = 2,\qquad (-2, -2, -1) \begin{pmatrix} -\dfrac{46}{15} \\ -\dfrac{14}{15} \\ -2 \end{pmatrix} = 10 .$$

Both are positive so the negative gradient itself is not an admissible direction. Thus we proceed to the computation of the dual basis for a_2, a_4. We have

$$\Gamma \begin{pmatrix} x_1 \\ y_1 \\ z_1 \end{pmatrix} = \begin{pmatrix} 0 & 0 & -1 \\ -2 & -2 & -1 \end{pmatrix}, \quad (\Gamma\Gamma')^{-1} = G^{-1} \begin{pmatrix} x_1 \\ y_1 \\ z_1 \end{pmatrix} = \begin{pmatrix} \frac{9}{8} & -\frac{1}{8} \\ -\frac{1}{8} & \frac{1}{8} \end{pmatrix},$$

$$(p_2, p_4) = D \begin{pmatrix} x_1 \\ y_1 \\ z_1 \end{pmatrix} = \Gamma' \begin{pmatrix} x_1 \\ y_1 \\ z_1 \end{pmatrix} G^{-1} \begin{pmatrix} x_1 \\ y_1 \\ z_1 \end{pmatrix} = \begin{pmatrix} \frac{1}{4} & -\frac{1}{4} \\ \frac{1}{4} & -\frac{1}{4} \\ -1 & 0 \end{pmatrix}.$$

Then

$$\mu^2 = (-\frac{46}{15}, -\frac{14}{15}, -2) \begin{pmatrix} \frac{1}{4} \\ \frac{1}{4} \\ -1 \end{pmatrix} = 1,$$

$$\mu^4 = (-\frac{46}{15}, -\frac{14}{15}, -2) \begin{pmatrix} -\frac{1}{4} \\ -\frac{1}{4} \\ 0 \end{pmatrix} = 1.$$

The projection of the negative gradient onto the tangent space is

$$\begin{pmatrix} -\dfrac{46}{15} \\ -\dfrac{14}{15} \\ -2 \end{pmatrix} - 1 \begin{pmatrix} 0 \\ 0 \\ -1 \end{pmatrix} - 1 \begin{pmatrix} -2 \\ -2 \\ -1 \end{pmatrix} = \begin{pmatrix} -\dfrac{16}{15} \\ \dfrac{16}{15} \\ 0 \end{pmatrix}$$

and this is equal to the new feasible direction r_1 since both μ_2 and μ_4 are positive.

We compute $\tilde{\lambda}$ again:

For $i = 1$, $(0, 0, 1) \begin{pmatrix} -\dfrac{16}{15} \\ \dfrac{16}{15} \\ 0 \end{pmatrix} = 0$, no restriction;

For $i = 3$, $(2, 2, 1) \begin{pmatrix} -\dfrac{16}{15} \\ \dfrac{16}{15} \\ 0 \end{pmatrix} = 0$, no restriction;

For $i = 5$, $(2, -1, -1) \begin{pmatrix} -\dfrac{16}{15} \\ \dfrac{16}{15} \\ 0 \end{pmatrix} = -\dfrac{48}{15}$, no restriction;

For $i = 6$, $\quad (-2, 1, 1) \begin{pmatrix} -\dfrac{16}{15} \\[2mm] \dfrac{16}{15} \\[2mm] 0 \end{pmatrix} = \dfrac{48}{15}$,

so we compute

$$- \dfrac{\left[(-2, 1, 1) \begin{pmatrix} -\dfrac{22}{15} \\[2mm] -\dfrac{38}{15} \\[2mm] -2 \end{pmatrix} - 8 \right]}{\dfrac{48}{15}} = 3 \ .$$

Thus $\tilde{\lambda} = 3$. To minimize $f \left(\begin{pmatrix} -\dfrac{22}{15} \\[2mm] -\dfrac{38}{15} \\[2mm] -2 \end{pmatrix} + \lambda \begin{pmatrix} -\dfrac{16}{15} \\[2mm] \dfrac{16}{15} \\[2mm] 0 \end{pmatrix} \right)$ we

compute

$$\dfrac{-2\left(-\dfrac{22}{15}, -\dfrac{38}{15}, -2\right) \begin{pmatrix} 1 & 0 & 0 \\ 0 & 1 & 0 \\ 0 & 0 & 1 \end{pmatrix} \begin{pmatrix} -\dfrac{16}{15} \\[2mm] \dfrac{16}{15} \\[2mm] 0 \end{pmatrix} - (6,6,6) \begin{pmatrix} -\dfrac{16}{15} \\[2mm] \dfrac{16}{15} \\[2mm] 0 \end{pmatrix}}{2\left(-\dfrac{16}{15}, \dfrac{16}{15}, 0\right) \begin{pmatrix} 1 & 0 & 0 \\ 0 & 1 & 0 \\ 0 & 0 & 1 \end{pmatrix} \begin{pmatrix} -\dfrac{16}{15} \\[2mm] \dfrac{16}{15} \\[2mm] 0 \end{pmatrix}} = \dfrac{1}{2} \ .$$

Thus $\lambda_1 = \min\{3, \frac{1}{2}\} = \frac{1}{2}$ and we obtain a new point

$$\begin{pmatrix} x_2 \\ y_2 \\ z_2 \end{pmatrix} = \begin{pmatrix} -\frac{22}{15} \\ -\frac{38}{15} \\ -2 \end{pmatrix} + \frac{1}{2}\begin{pmatrix} -\frac{16}{15} \\ \frac{16}{15} \\ 0 \end{pmatrix} = \begin{pmatrix} -2 \\ -2 \\ -2 \end{pmatrix}.$$

<u>Step 3.</u> Because $\lambda_1 < \tilde{\lambda}$ and because the feasible direction in Step 2 was the projection of the negative gradient on the tangent space the active constraints remain unchanged. (This may be verified by direct computation, of course.) The matrices Γ and G and the dual basis p_2, p_4 remain unchanged. The new value of the negative gradient is

$$(-2(-2+3), \quad -2(-2+3), \quad -2(-2+3)) = (-2, -2, -2).$$

We then use p_2 and p_4, already found in Step 2 to compute

$$\mu^2 = (-2, -2, -2)\begin{pmatrix} \frac{1}{4} \\ \frac{1}{4} \\ -1 \end{pmatrix} = 1,$$

$$\mu^4 = (-2, -2, -2) \begin{pmatrix} -\dfrac{1}{4} \\ -\dfrac{1}{4} \\ 0 \end{pmatrix} = 1 \ .$$

The projection of the negative gradient on the tangent space is

$$\begin{pmatrix} -2 \\ -2 \\ -2 \end{pmatrix} - 1 \begin{pmatrix} 0 \\ 0 \\ -1 \end{pmatrix} - 1 \begin{pmatrix} -2 \\ -2 \\ -1 \end{pmatrix} = \begin{pmatrix} 0 \\ 0 \\ 0 \end{pmatrix} \ .$$

Thus the Kuhn Tucker conditions are satisfied at the point

$\begin{pmatrix} -2 \\ -2 \\ -2 \end{pmatrix}$. Since the objective function is convex, Theorem 10.1

tells us that we have indeed reaches the minimum for f in S

and we stop.

EXAMPLE. $h(x, y, z) = (y - 10)^2$.

This time we obtain an initial point in S:

$$\begin{pmatrix} x_0 \\ y_0 \\ z_0 \end{pmatrix} = \begin{pmatrix} -4 \\ -2 \\ 2 \end{pmatrix}$$

by treating the first, fourth and sixth constraints as equalities. It should be noted here that this procedure for finding an initial point in S works only because our set S is so very special. Even for polyhedral sets S, if the number of inequalities used to define S is large it may be very difficult to find a suitable initial point in S. Special techniques have been devised for finding such a point but these are outside the scope of this book. Often in the case of real-life problems a suitable initial point can be obtained by inspection – not always, however.

Step 1. The active constraints at the initial point are specified

by $I \begin{pmatrix} x_0 \\ y_0 \\ z_0 \end{pmatrix} = \{1, 4, 6\}$. The negative gradient is

$$(0, -2(-2 - 10), 0) = (0, 24, 0).$$

$$\Gamma \begin{pmatrix} -4 \\ -2 \\ 2 \end{pmatrix} = \begin{pmatrix} 0 & 0 & 1 \\ -2 & -2 & -1 \\ -2 & 1 & 1 \end{pmatrix},$$

$$
D \begin{pmatrix} -4 \\ -2 \\ 2 \end{pmatrix} = \Gamma^{-1} \begin{pmatrix} -4 \\ -2 \\ 2 \end{pmatrix} = \begin{pmatrix} \frac{1}{6} & -\frac{1}{6} & -\frac{1}{3} \\ -\frac{2}{3} & -\frac{1}{3} & \frac{1}{3} \\ 1 & 0 & 0 \end{pmatrix} = (p_1, p_4, p_6).
$$

The negative gradient itself proves not to be a feasible direction so we compute

$$
\mu^1 = (0, 24, 0) \begin{pmatrix} \frac{1}{6} \\ -\frac{2}{3} \\ 1 \end{pmatrix} = -16
$$

$$
\mu^4 = (0, 24, 0) \begin{pmatrix} -\frac{1}{6} \\ -\frac{1}{3} \\ 0 \end{pmatrix} = -8
$$

$$
\mu^6 = (0, 24, 0) \begin{pmatrix} -\frac{1}{3} \\ \frac{1}{3} \\ 0 \end{pmatrix} = 8 .
$$

Thus

$$
r_0 = \mu^1 p_1 + \mu^4 p_4 = -16 \begin{pmatrix} \frac{1}{6} \\ -\frac{2}{3} \\ 1 \end{pmatrix} - 8 \begin{pmatrix} -\frac{1}{6} \\ -\frac{1}{3} \\ 0 \end{pmatrix} = \begin{pmatrix} -\frac{4}{3} \\ \frac{40}{3} \\ -16 \end{pmatrix}.
$$

(Since a_1, a_2, a_4 span R^3 the tangent space is null and we do not need to compute the projection of the negative gradient on it.) To obtain $\tilde{\lambda}$ we compute

For $i = 2$
$$\frac{- \left[(0, 0, -1) \begin{pmatrix} -4 \\ -2 \\ 2 \end{pmatrix} - 2 \right]}{(0, 0, -1) \begin{pmatrix} -\frac{4}{3} \\ \frac{40}{3} \\ -16 \end{pmatrix}} = \frac{1}{4} \; ;$$

For $i = 3$,
$$\frac{- \left[(2, 2, 1) \begin{pmatrix} -4 \\ -2 \\ 2 \end{pmatrix} - 10 \right]}{(2, 2, 1) \begin{pmatrix} -\frac{4}{3} \\ -\frac{40}{3} \\ -16 \end{pmatrix}} = \frac{5}{2} \; ;$$

For $i = 5$, $(2, -1, -1) \begin{pmatrix} -\frac{4}{3} \\ \frac{40}{3} \\ -16 \end{pmatrix} = 0$,   no restriction.

Thus $\tilde{\lambda} = \min \{ \frac{1}{4}, \frac{5}{2} \} = \frac{1}{4}$.   To minimize $h \left(\begin{pmatrix} -4 \\ -2 \\ 2 \end{pmatrix} + \lambda \begin{pmatrix} -\frac{4}{3} \\ -2 \\ 2 \end{pmatrix} \right)$

we compute

$$\frac{-2(-4,-2,2)\begin{pmatrix} 0 & 0 & 0 \\ 0 & 1 & 0 \\ 0 & 0 & 0 \end{pmatrix}\begin{pmatrix} -\dfrac{4}{3} \\ \dfrac{40}{3} \\ -16 \end{pmatrix} + (0,20,0)\begin{pmatrix} -\dfrac{4}{3} \\ \dfrac{40}{3} \\ -16 \end{pmatrix}}{2(-\dfrac{4}{3},\dfrac{40}{3},-16)\begin{pmatrix} 0 & 0 & 0 \\ 0 & 1 & 0 \\ 0 & 0 & 0 \end{pmatrix}\begin{pmatrix} -\dfrac{4}{3} \\ \dfrac{40}{3} \\ -16 \end{pmatrix}} = \frac{9}{10} .$$

Thus $\lambda_0 = \min\{\frac{1}{4}, \frac{9}{10}\} = \frac{1}{4}$ and our new point is

$$\begin{pmatrix} x_1 \\ y_1 \\ z_1 \end{pmatrix} = \begin{pmatrix} -4 \\ -2 \\ 2 \end{pmatrix} + \frac{1}{4}\begin{pmatrix} -\dfrac{4}{3} \\ \dfrac{40}{3} \\ -16 \end{pmatrix} = \begin{pmatrix} -\dfrac{13}{3} \\ \dfrac{4}{3} \\ -2 \end{pmatrix} .$$

<u>Step 2.</u> We readily verify $I\begin{pmatrix} -\dfrac{13}{3} \\ \dfrac{4}{3} \\ -2 \end{pmatrix} = \{2, 6\}.$

The negative gradient is

$$(0, -2(\frac{4}{3} - 10), 0) = (0, \frac{52}{3}, 0).$$

We have

$$\Gamma \begin{pmatrix} -\dfrac{13}{3} \\ \dfrac{4}{3} \\ -2 \end{pmatrix} = \begin{pmatrix} 0 & 0 & -1 \\ & & \\ -2 & 1 & 1 \end{pmatrix}, \quad G^{-1}\begin{pmatrix} -\dfrac{13}{3} \\ \dfrac{4}{3} \\ -2 \end{pmatrix} = \begin{pmatrix} \dfrac{6}{5} & \dfrac{1}{5} \\ \dfrac{1}{5} & \dfrac{1}{5} \end{pmatrix},$$

$$(p_2, p_6) = D\begin{pmatrix} -\dfrac{13}{3} \\ \dfrac{4}{3} \\ -2 \end{pmatrix} = \Gamma'\begin{pmatrix} -\dfrac{13}{3} \\ \dfrac{4}{3} \\ -2 \end{pmatrix}, G^{-1}\begin{pmatrix} -\dfrac{13}{3} \\ \dfrac{4}{3} \\ -2 \end{pmatrix} = \begin{pmatrix} -\dfrac{2}{5} & -\dfrac{2}{5} \\ \dfrac{1}{5} & \dfrac{1}{5} \\ -1 & 0 \end{pmatrix}.$$

Again the negative gradient direction is not itself feasible as one may readily check. So we compute

$$\mu^2 = (0, \frac{52}{3}, 0)\begin{pmatrix} -\dfrac{2}{5} \\ \dfrac{1}{5} \\ -1 \end{pmatrix} = \frac{52}{15} \,,$$

$$\mu^6 = (0, \frac{52}{3}, 0)\begin{pmatrix} -\dfrac{2}{5} \\ \dfrac{1}{5} \\ 0 \end{pmatrix} = \frac{52}{15} \,.$$

We compute the projection of the negative gradient on the tangent space:

$$\begin{pmatrix} 0 \\ \dfrac{52}{3} \\ 0 \end{pmatrix} - \dfrac{52}{15} \begin{pmatrix} 0 \\ 0 \\ -1 \end{pmatrix} - \dfrac{52}{15} \begin{pmatrix} -2 \\ 1 \\ 1 \end{pmatrix} = \begin{pmatrix} \dfrac{104}{15} \\ \dfrac{208}{15} \\ 0 \end{pmatrix} = r_1 \, ,$$

the new feasible direction, because μ^2 and μ^6 are both positive. Next we compute $\tilde{\lambda}$:

For $i = 1$, $\quad (0,0,1) \begin{pmatrix} \dfrac{104}{15} \\ \dfrac{208}{15} \\ 0 \end{pmatrix} = 0$, $\quad$ no restriction;

For $i = 3$, $\quad \dfrac{- \left[(2,2,1) \begin{pmatrix} -\dfrac{13}{3} \\ \dfrac{4}{3} \\ -2 \end{pmatrix} - 10 \right]}{(2,2,1) \begin{pmatrix} \dfrac{104}{15} \\ \dfrac{208}{15} \\ 0 \end{pmatrix}} = \dfrac{45}{104}$;

For $i = 4$, $\quad (-2,-2,-1) \begin{pmatrix} \dfrac{104}{15} \\ \dfrac{208}{15} \\ 0 \end{pmatrix} = -\dfrac{208}{5} < 0$, no restriction;

For $i = 5$, $(2, -1, -1) \begin{pmatrix} \dfrac{104}{15} \\ \dfrac{208}{15} \\ 0 \end{pmatrix} = 0$,   no restriction.

Thus $\tilde{\lambda} = \dfrac{45}{104}$.   To minimize $h\left(\begin{pmatrix} -\dfrac{13}{3} \\ \dfrac{4}{3} \\ -2 \end{pmatrix} + \lambda \begin{pmatrix} \dfrac{104}{15} \\ \dfrac{208}{15} \\ 0 \end{pmatrix} \right)$ we

compute

$$\frac{-2\left(-\dfrac{13}{3}, \dfrac{4}{3}, -2\right) \begin{pmatrix} 0 & 0 & 0 \\ 0 & 1 & 0 \\ 0 & 0 & 0 \end{pmatrix} \begin{pmatrix} \dfrac{104}{15} \\ \dfrac{208}{15} \\ 0 \end{pmatrix} + (0,20,0) \begin{pmatrix} \dfrac{104}{15} \\ \dfrac{208}{15} \\ 0 \end{pmatrix}}{2\left(\dfrac{104}{15}, \dfrac{208}{15}, 0\right) \begin{pmatrix} 0 & 0 & 0 \\ 0 & 1 & 0 \\ 0 & 0 & 0 \end{pmatrix} \begin{pmatrix} \dfrac{104}{15} \\ \dfrac{208}{15} \\ 0 \end{pmatrix}} = \frac{5}{8}.$$

Thus $\hat{\lambda} = \min\left\{\dfrac{45}{104}, \dfrac{5}{8}\right\} = \dfrac{45}{104}$ and our new point is

$$\begin{pmatrix} x_2 \\ y_2 \\ z_2 \end{pmatrix} = \begin{pmatrix} -\dfrac{13}{3} \\ \dfrac{4}{3} \\ -2 \end{pmatrix} + \frac{45}{104} \begin{pmatrix} \dfrac{104}{15} \\ \dfrac{208}{15} \\ 0 \end{pmatrix} = \begin{pmatrix} -\dfrac{4}{3} \\ \dfrac{22}{3} \\ -2 \end{pmatrix}.$$

<u>Step 3.</u>   The new active constraints are prescribed by

$$I \begin{pmatrix} -\dfrac{4}{3} \\ \dfrac{22}{3} \\ -2 \end{pmatrix} = \{2,\, 3,\, 6\}. \quad \text{Thus}$$

$$\Gamma \begin{pmatrix} -\dfrac{4}{3} \\ \dfrac{22}{3} \\ -2 \end{pmatrix} = \begin{pmatrix} 0 & 0 & -1 \\ 2 & 2 & 1 \\ -2 & 1 & 1 \end{pmatrix}, \quad (p_2, p_3, p_6) = D \begin{pmatrix} -\dfrac{4}{3} \\ \dfrac{22}{3} \\ -2 \end{pmatrix}$$

$$= \Gamma^{-1} \begin{pmatrix} -\dfrac{4}{3} \\ \dfrac{22}{3} \\ -2 \end{pmatrix} = \begin{pmatrix} -\dfrac{1}{6} & \dfrac{1}{6} & -\dfrac{1}{3} \\ \dfrac{2}{3} & \dfrac{1}{3} & \dfrac{1}{3} \\ -1 & 0 & 0 \end{pmatrix} \; .$$

The negative gradient is

$$(0,\, -2(\tfrac{22}{3} - 10),\, 0) = (0,\, \tfrac{16}{3},\, 0).$$

We compute μ^2, μ^3, μ^6:

$$\mu^2 = (0,\, \tfrac{16}{3},\, 0) \begin{pmatrix} -\dfrac{1}{6} \\ \dfrac{2}{3} \\ -1 \end{pmatrix} = \dfrac{32}{9},$$

$$\mu^3 = (0, \frac{16}{3}, 0) \begin{pmatrix} \frac{1}{6} \\ \frac{1}{3} \\ 0 \end{pmatrix} = \frac{16}{9} ,$$

$$\mu^6 = (0, \frac{16}{3}, 0) \begin{pmatrix} -\frac{1}{3} \\ \frac{1}{3} \\ 0 \end{pmatrix} = \frac{16}{9} .$$

Since all three of these are positive and the tangent space is null the Kuhn-Tucker conditions are satisfied. Theorem 10.2 then shows that $x = -\frac{4}{3}$, $y = \frac{22}{3}$, $z = -2$ is the solution of our problem. Thus the treatment of this example is complete.

The reader should note that it is entirely possible for the minimizing x^* to lie in the interior of the set S in some cases. When this is true the point x^* will be approached by a sequence of points x_k, all but finitely many of them lying in the interior of S where no constraints are active. At such points x_k where no constraints are active the feasible direction $r_k = \frac{\partial f}{\partial x}(x_k)'$ is always selected. We select λ_k as in the other cases. In fact this method of continuing along in the direction $r_k = -\frac{\partial f}{\partial x}(x_k)'$ until $f(x_k - \lambda r_k)$ is minimized provides a possible alternative to the gradient method discussed in Chapter 8 for unconstrained problems.

It is not too difficult to provide a proof for the convergence of the gradient projection technique for linear inequality constraints which we have described in this chapter. We will not present such a proof in this text, however.

When the objective function is linear:

$$f(x) = a'x + b$$

(or, equivalently, $f(x) = a'x$, since b is a constant), and the set S is described by linear inequalities as above, we have what is known as a linear programming problem. A special technique, known as the <u>Simplex Method</u>, handles such problems very efficiently. We will not discuss that method in this text because it has been adequately treated in many others. However, gradient projection can also be used to solve such problems fairly efficiently. The method proceeds just as we have described it for general objective functions but there are a number of simplifications. First of all, $\frac{\partial f}{\partial x} \equiv a'$ and thus need not be recomputed at every step. It is easy to see that the minimum x^* lies on the boundary of S if $a \neq 0$ and it can be shown that x^* will be achieved in a finite number of steps using the gradient projection technique.

The greatest simplification arises from the fact that the auxiliary minimization of $f(x_k + \lambda r_k)$ is trivial. For if r_k is a feasible direction at x_k then $\frac{\partial f}{\partial x}(x_k)r_k = a'r_k < 0$. But

$$\frac{d}{d\lambda} f(x_k + \lambda r_k) = \frac{d}{d\lambda}(a'(x_k + \lambda r_k) + b) = a'r_k < 0$$

for all values of λ. It follows that $f(x_k + \lambda r_k)$ cannot achieve a minimum in the interior of the interval $[0, \tilde{\lambda}]$. We therefore always take $\lambda = \tilde{\lambda} = \min_{\substack{i \in I(x_k) \\ a_i'r_k > 0}} \left\{ \frac{-[a_i'x_k + b_i]}{a_i'r_k} \right\}$ if this quantity is defined. If there are no a_i with $a_i'r_k > 0$ then the linear programming problem has no solution (or has the solution $f = -\infty$, if we prefer). We see then that the solution of linear programming problems is an entirely algebraic finite process wherein no nonterminating iterative procedures are required.

We conclude this chapter with a short discussion of

the problem of degeneracy and how we find a feasible direction $\hat{r}$, or else determine that the Kuhn Tucker conditions are satisfied, when $\hat{x}$ is a point of degeneracy. Recall that $\hat{x}$ is a point of degeneracy if the vectors a_i, $i \in I(\hat{x})$, are not linearly independent, i. e. , if the rank of the matrix $\Gamma(\hat{x})$ is less than $m(\hat{x})$, the number of active constraints at $\hat{x}$. This rank deficiency will be betrayed by the impossibility of inverting the matrix $G(\hat{x}) = \Gamma(\hat{x})\Gamma(\hat{x})'$. The actual rank of $\Gamma(\hat{x})$ can be determined by the process of Gaussian elimination which is described in most numerical analysis texts and is the usual method used to solve linear equations.

Degeneracy will definitely occur if $m(\hat{x}) > n$, where n is the dimension of the space in which we are working. As an example of this first situation we consider the set S_1 shown in Figure 23. Its description is

$$S_1 = \left\{ \begin{pmatrix} x \\ y \\ z \end{pmatrix} \in R^3 \,\middle|\, -z \leq 0, \; x+y+z-1 \leq 0, \right.$$

$$2x - y + 2z - 2 \leq 0, \; -x - 2y + 3z - 3 \leq 0,$$

$$\left. z - 1 \leq 0, \quad y \leq 0 \right\} .$$

Thus

$$a_1 = \begin{pmatrix} 0 \\ 0 \\ -1 \end{pmatrix}, \ a_2 = \begin{pmatrix} 1 \\ 1 \\ 1 \end{pmatrix}, \ a_3 = \begin{pmatrix} 2 \\ -1 \\ 2 \end{pmatrix}, \ a_4 = \begin{pmatrix} -1 \\ -2 \\ 3 \end{pmatrix},$$

$$a_5 = \begin{pmatrix} 0 \\ 0 \\ 1 \end{pmatrix}, \ a_6 = \begin{pmatrix} 0 \\ 1 \\ 0 \end{pmatrix}.$$

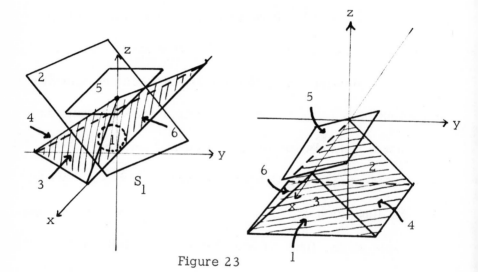

Figure 23

At the point $\begin{pmatrix} 0 \\ 0 \\ 1 \end{pmatrix}$ the last five constraints are all active, i. e.
$I(\begin{smallmatrix} 0 \\ 0 \\ 1 \end{smallmatrix}) = \{2, 3, 4, 5, 6\}$. Since the dimension of the space we
are working in is three and five constraints are active, the
point $\begin{pmatrix} 0 \\ 0 \\ 1 \end{pmatrix}$ is a point of degeneracy.

Consider also the set S_2 shown in Figure 23, described

by

$$S_2 = \left\{ \begin{pmatrix} x \\ y \\ z \end{pmatrix} \in R^3 \, | \, x - 1 \le 0, \ -x \le 0, \ -z - 1 \le 0, \right.$$

$$\left. y + z \le 0, \ -y + 2z \le 0, \ -y + z \le 0 \right\} .$$

Here

$$a_1 = \begin{pmatrix} 1 \\ 0 \\ 0 \end{pmatrix}, \quad a_2 = \begin{pmatrix} -1 \\ 0 \\ 0 \end{pmatrix}, \quad a_3 = \begin{pmatrix} 0 \\ 0 \\ -1 \end{pmatrix}, \quad a_4 = \begin{pmatrix} 0 \\ 1 \\ 1 \end{pmatrix},$$

$$a_5 = \begin{pmatrix} 0 \\ -1 \\ 2 \end{pmatrix}, \quad a_6 = \begin{pmatrix} 0 \\ -1 \\ 1 \end{pmatrix} .$$

At the point $\begin{pmatrix} \frac{1}{2} \\ 0 \\ 0 \end{pmatrix}$ the last three constraints are active so

$I \begin{pmatrix} \frac{1}{2} \\ 0 \\ 0 \end{pmatrix} = \{4, 5, 6\}$. But a_4, a_5, a_6 are clearly not linearly in-

dependent so once again we have a point of degeneracy. In

fact, all points $\begin{pmatrix} x \\ 0 \\ 0 \end{pmatrix}$, $0 \le x \le 1$, are points of degeneracy in

this case.

When $\hat{x}$ is a point of degeneracy the matrix $G(\hat{x})^{-1}$

does not exist and we cannot proceed in the usual way to find

a feasible direction $\hat{r}$. We shall now describe a method for

finding a feasible direction $\hat{r}$ which will apply in most cases of degeneracy. We make a special assumption, namely the

GENERAL POSITION ASSUMPTION. Let $\tilde{m}$ = rank $\Gamma(\hat{x})$, i. e.,

the number of elements in a maximal linearly independent sub-

set of $\{a_i \mid i \in I(\hat{x})\}$. Then every subset of $\{a_i \mid i \in I(\hat{x})\}$

consisting of $\tilde{m}$ or fewer vectors is linearly independent.

This general position assumption will almost always

be valid in practice, though one can easily construct "text-

book" examples where it fails.

Let $\hat{x}$ be a point of degeneracy satisfying the general

position condition. Let $I(\hat{x})$ be the set of indices corres-

ponding to the active constraints at $\hat{x}$ and let $\tilde{m}$ = rank $\Gamma(\hat{x})$.

We proceed as follows to find a feasible direction r:

(i)   Compute $-\frac{\partial f}{\partial x}(\hat{x})a_i$, $i \in I(\hat{x})$. If this quantity is ≤ 0

for all such i we put $\hat{r} = -\frac{\partial f}{\partial x}(\hat{x})'$. Otherwise, proceed to

(ii).

(ii) Let $I(\hat{x}) = \tilde{I} \cup \tilde{J}$, where $\tilde{I}$ is a subset consisting of $\tilde{m}$

(or fewer) of the indices in $I(\hat{x})$ and $\tilde{J}$ consists of the re-

maining indices. Let $\tilde{\Gamma}$ be the $\tilde{m}$ by n matrix whose rows

are a_i', $i \in \tilde{I}$. Compute a direction $\tilde{r}$ as in the case of a regular point but using $\tilde{I}$ and $\tilde{\Gamma}$ instead of $I(\hat{x})$ and $\Gamma(\hat{x})$.

That is

$$\tilde{D} = \tilde{\Gamma}'(\tilde{\Gamma}\tilde{\Gamma}')^{-1} = \tilde{\Gamma}'\tilde{G}^{-1}.$$

The columns of $\tilde{D}$ are a dual basis, p_i, $i \in \tilde{I}$, for the subspace spanned by a_i, $i \in \tilde{I}$. Then

$$\mu^i = -\frac{\partial f}{\partial x}(\hat{x})p_i, \quad i \in \tilde{I}.$$

Put

$$\tilde{z} = -\frac{\partial f}{\partial x}(\hat{x})' - \sum_{i \in \tilde{I}} \mu^i a_i,$$

$$\tilde{r} = \sum_{\substack{i \in \tilde{I} \\ \mu^i < 0}} \mu^i p_i + \tilde{z}.$$

If $\tilde{z} = 0$ and all $\mu^i \geq 0$, $i \in \tilde{I}$, then $\tilde{r} = 0$ and the Kuhn-Tucker conditions are satisfied by taking $\mu^i = 0$, $i \in \tilde{J}$. In this case we stop since x may be a solution to our minimization problem. (It definitely is if the conditions of Theorem 10.2 hold.) If $\tilde{r} \neq 0$ we compute

$$a_i'r, \quad i \in \tilde{J}.$$

If all of these numbers are ≤ 0 we put $\hat{r} = \tilde{r}$ and we have

the desired feasible direction.   Otherwise we proceed to (iii).

(iii)   We let $\tilde{I} = I_1 \cup I_2$, $\tilde{J} = J_1 \cup J_2$, where

$$\mu^i \geq 0, \ i \in I_1; \quad \mu^i < 0, \quad i \in I_2;$$

$$a_i'\tilde{r} > 0, \ i \in J_1; \quad a_i'\tilde{r} \leq 0, \quad i \in J_2.$$

We form a new set $\tilde{\tilde{I}}$ of $\tilde{m}$ or fewer indices by adjoining to

I_1 some or all of the vectors in J_1 and then return to (ii),

replacing $\tilde{I}$ by $\tilde{\tilde{I}}$ and $\tilde{J}$ by $\tilde{\tilde{J}} = I(\hat{x}) - \tilde{\tilde{I}}$.

The process outlined above eventually terminates by

providing us with a feasible direction or by showing us that

the Kuhn-Tucker conditions are satisfied.

To provide an example we let

$$f(x, y, z) = 5x + 2y - z$$

and take S to be the set S_1 described above.   We start at

the point of degeneracy $\hat{x} = 0$, $\hat{y} = 0$, $\hat{z} = 1$.   We have

$$I \begin{pmatrix} 0 \\ 0 \\ 1 \end{pmatrix} = \{2, 3, 4, 5, 6\}.$$

Let us put $\tilde{I} = \{4, 5, 6\}$. Then

$$\tilde{\Gamma} = \begin{pmatrix} -1 & -2 & 3 \\ 0 & 0 & 1 \\ 0 & 1 & 0 \end{pmatrix} \quad \tilde{D} = (\tilde{p}_4, \tilde{p}_5, \tilde{p}_6) = \tilde{\Gamma}^{-1} = \begin{pmatrix} -1 & 3 & -2 \\ 0 & 0 & 1 \\ 0 & 1 & 0 \end{pmatrix}.$$

The negative gradient is $(-5, -2, 1)$ and we compute

$$\tilde{\mu}^4 = (-5, -2, 1) \begin{pmatrix} -1 \\ 0 \\ 0 \end{pmatrix} = 5, \quad \tilde{\mu}^5 = (-5, -2, 1) \begin{pmatrix} 3 \\ 0 \\ 1 \end{pmatrix} = -14,$$

$$\tilde{\mu}^6 = (-5, -2, 1) \begin{pmatrix} -2 \\ 1 \\ 0 \end{pmatrix} = 8 .$$

Thus we put $\tilde{r} = \tilde{\mu}^5 \tilde{p}_5 = -14 \begin{pmatrix} 3 \\ 0 \\ 1 \end{pmatrix} = \begin{pmatrix} -42 \\ 0 \\ -14 \end{pmatrix}$. Moreover,

$I_1 = \{4, 6\}, \quad I_2 = \{5\}$. Now

$$a_2'\tilde{r} = (1, 1, 1) \begin{pmatrix} -42 \\ 0 \\ -14 \end{pmatrix} = -56, \quad a_3'\tilde{r} = (2, -1, 2) \begin{pmatrix} -42 \\ 0 \\ -14 \end{pmatrix} = -112.$$

Therefore $\tilde{r}$ is a feasible direction and could be used to leave

the point $\hat{x} = 0, \hat{y} = 0, \hat{z} = 1$.

Now let us try again with

$$h(x, y, z) = -5x - 2y + z = -f(x, y, z).$$

If we start with $\tilde{I} = \{4,5,6\}$ it is clear that we will have

$\tilde{\mu}^4 = -5$, $\tilde{\mu}^5 = 14$, $\tilde{\mu}^6 = -8$ so we use

$$\tilde{r} = \tilde{\mu}^4 \tilde{p}_4 + \tilde{\mu}^6 \tilde{p}_6 = -5 \begin{pmatrix} -1 \\ 0 \\ 0 \end{pmatrix} - 8 \begin{pmatrix} -2 \\ 1 \\ 0 \end{pmatrix} = \begin{pmatrix} 21 \\ -8 \\ 0 \end{pmatrix}.$$

Moreover $I_1 = \{5\}$, $I_2 = \{4,6\}$. We compute

$$a_2' \tilde{r} = (1,1,1) \begin{pmatrix} 21 \\ -8 \\ 0 \end{pmatrix} = 13, \quad a_3' \tilde{r} = (2,-1,2) \begin{pmatrix} 21 \\ -8 \\ 0 \end{pmatrix} = 50.$$

Thus $J_1 = \{2,3\}$ and J_2 is empty. We put $\tilde{\tilde{I}} = I_1 \cup J_1 = \{2,3,5\}$ and again start the process described in (ii).

$$\tilde{\tilde{\Gamma}} = \begin{pmatrix} 1 & 1 & 1 \\ 2 & -1 & 2 \\ 0 & 0 & 1 \end{pmatrix}, \quad \tilde{\tilde{D}} = (\tilde{\tilde{p}}_2, \tilde{\tilde{p}}_3, \tilde{\tilde{p}}_5) = \tilde{\tilde{\Gamma}}^{-1} \begin{pmatrix} \frac{1}{3} & \frac{1}{3} & -1 \\ \frac{2}{3} & -\frac{1}{3} & 0 \\ 0 & 0 & 1 \end{pmatrix},$$

$$\tilde{\tilde{\mu}}^2 = (5,2,-1) \begin{pmatrix} \frac{1}{3} \\ \frac{2}{3} \\ 0 \end{pmatrix} = 3, \quad \tilde{\tilde{\mu}}^3 = (5,2,-1) \begin{pmatrix} \frac{1}{3} \\ -\frac{1}{3} \\ 0 \end{pmatrix} = 1,$$

$$\tilde{\tilde{\mu}}^5 = (5,2,-1) \begin{pmatrix} -1 \\ 0 \\ 1 \end{pmatrix} = -6.$$

We put

$$\overset{\approx}{r} = \overset{\approx 5}{\mu} \overset{\approx 5}{p} = -6 \begin{pmatrix} -1 \\ 0 \\ 1 \end{pmatrix} = \begin{pmatrix} 6 \\ 0 \\ -6 \end{pmatrix} .$$

Now

$$a'_4 \overset{\approx}{r} = (-1, -2, 3) \begin{pmatrix} 6 \\ 0 \\ -6 \end{pmatrix} = -24 , \quad a'_6 \overset{\approx}{r} = (0, 1, 0) \begin{pmatrix} 6 \\ 0 \\ -6 \end{pmatrix} = 0.$$

Thus $\overset{\approx}{r}$ is a feasible direction and we put $r = \overset{\approx}{r} = \begin{pmatrix} 6 \\ 0 \\ -6 \end{pmatrix}$ to

leave the point $\hat{x} = 0$, $\hat{y} = 0$, $\hat{z} = 1$.

EXERCISES. CHAPTER 10

1. Suppose that at a certain point in R^4 the gradients of the

active constraints are

$$a'_1 = (4, 3, 2, 1), \quad a'_2 = (3, 2, 1, 0), \quad a'_3 = (2, 1, 0, 0)$$

and assume that the negative gradient at this point is

$(-1, 1, 3, -2)$. Compute z, p_1, p_2, p_3, μ^1, μ^2, μ^3 and

determine either a feasible direction or that the Kuhn–

Tucker conditions are satisfied.

2. Let $S = \{x \in R^n \mid a'_i x + b_i \leq 0, \; i = 1, 2, \ldots, m\}$. Let

$x^*\, \epsilon\, S$ and assume the Kuhn-Tucker conditions

$$\frac{\partial f}{\partial x}(\hat{x}) + \sum_{i\,\epsilon\, I(x^*)} \mu^i a_i' = 0, \quad \mu^i \geq 0, \quad i\,\epsilon\, I(x^*)$$

are satisfied at x^*. Show that there is no point $x\,\epsilon\, S$ such that $\dfrac{\partial f}{\partial x}(x^*)(x - x^*) < 0$.

3.   (Suitable for either hand computation or a computer assignment.) Let $S = \{(\begin{smallmatrix}x\\y\\z\end{smallmatrix})\,\epsilon\, R^3 \,|\, a_i'x + b_i \leq 0, \; i = 1,\ldots,5\}$, where

$$a_1 = \begin{pmatrix} 1 \\ 1 \\ 1 \end{pmatrix}, \quad a_2 = \begin{pmatrix} -2 \\ 1 \\ 1 \end{pmatrix}, \quad a_3 = \begin{pmatrix} -1 \\ -2 \\ 2 \end{pmatrix}, \quad a_4 = \begin{pmatrix} 2 \\ -1 \\ -2 \end{pmatrix},$$

$$a_5 = \begin{pmatrix} -1 \\ -1 \\ -3 \end{pmatrix},$$

$$b_1 = -4, \quad b_2 = -5, \quad b_3 = -6, \quad b_4 = -7, \quad b_5 = -8.$$

Use the gradient projection technique to find the minimum for

$$f(x, y, z) = x^3 - 2x^2 + x + xy + y^3 - 4y^2 + 4y + z^3 + z^2 y \ .$$

Note the possibility of local minima which are not global minima.

4. Because of a prolonged strike in the steel industry a certain furniture factory is left with limited supplies of hardware. In inventory are the following quantities

nails	14, 000
bolts	1, 200
staples	8, 500
screws	2, 600
drawer knobs	320
hinges	100

The factory produces the following items each of which requires hardware as indicated by the table below. Also indicated in the table is the profit realized from each finished item.

	nails	bolts	staples	screws	knobs	hinges	profit
tables	28	20	16	12	0	0	$ 2.00
chairs	46	4	12	20	0	0	$ 1.50
desks	100	16	32	36	6	2	$ 5.00
cabinets	120	14	86	28	8	0	$ 7.50
chests of drawers	150	22	100	42	8	4	$10.00

A production plan calling for the production of x^1 tables, x^2 chairs, x^3 desks, x^4 cabinets and x^5 chests of drawers is to be drawn up in such a way that profit is maximized but hardware does not run short. Phrase this problem as a linear programming problem and write a computer program to solve it using the gradient projection method. In addition to the constraints which arise from the hardware limitations, do not forget $x^i \geq 0$, $i = 1, 2, 3, 4, 5$.

5.   Devise a technique of gradient projection type which will solve optimization problems subject to both linear equality and linear inequality constraints, i.e., where the feasible set S is of the form

$$S = \{x \in R^n \, | \, a_i'x + b_i \leq 0, \ i = 1, 2, \ldots, m,$$

$$c_j'x + d_j = 0, \ j = 1, \ldots, p\}.$$

Show that the appropriate analog of the Lagrange-Multiplier, Kuhn-Tucker conditions in this case is

$$\frac{\partial f}{\partial x}(x^*) + \sum_{i \in I(x^*)} \mu^i a_i' + \sum_{j=1}^{p} v^j c_j', \quad \mu^i \geq 0, \ v^j \text{ real}.$$

Use your procedure to solve problem 6, Chapter 9 under the additional constraints $|u_k| \leq \frac{1}{2}$, $k = 1, 2, 3, 4, 5$. (This is a relatively difficult problem and should be assigned only to very well prepared students).

6. Let S_1 and S_2 be the sets described in the foregoing chapter in connection with our discussion of the degeneracy problem. Solve the following linear programming problems by hand:

$$\text{maximize } x - 2y + z, \quad \begin{pmatrix} x \\ y \\ z \end{pmatrix} \in S_2. \quad \text{Start at } \begin{pmatrix} 0 \\ 0 \\ -1 \end{pmatrix}.$$

$$\text{maximize } x + 2y + 3z, \quad \begin{pmatrix} x \\ y \\ z \end{pmatrix} \in S_1. \quad \text{Start at } \begin{pmatrix} 0 \\ 0 \\ 0 \end{pmatrix}.$$

7. Let $g_i(x)$, $i = 1, 2, \ldots, m$ be functions as described in Chapter 9, such that the active constraints at any given point satisfy the algebraic constraint qualification. The functions $g_i(x)$ are in general nonlinear. Let $S = \{x \in R^n | g_i(x) \leq 0, i = 1, 2, \ldots, r\}$. Discuss how one might adapt the gradient projection to solve problems of the type

$$\text{minimize } f(x), \quad x \in S.$$

This will involve a combination of the methods of

Chapters 9 and 10. Can you state and prove analogs of

Theorems 10.1 and 10.2 for such problems?

8.   Let $a_1, a_2, \ldots, a_m$ be $m \leq n$ linearly independent vec-

tors in R^n. Let $p_1, p_2, \ldots, p_m$ be the dual basis for

the subspace

$$N = \{y \in R^n \,|\, y = \sum_{i=1}^{m} \alpha^i a_i, \quad \alpha^i \text{ real}\}$$

relative to $a_1, a_2, \ldots, a_m$. Define also

$$T = \{z \in R^n \,|\, y'z = 0, \quad y \in N\}.$$

Show that a vector $x \in R^n$ satisfies $a_i' x \leq 0$, $i = 1, 2, \ldots, m$

if and only if there is a vector $z \in T$ and real numbers

$\beta^i \leq 0$, $i = 1, 2, \ldots, m$, such that $x = z + \sum_{i=1}^{m} \beta^i p_i$. Sketch

the set $S = \{x \in R^3 \,|\, a_i' x \leq 0, \, i = 1, 2, 3\}$ with

$$a_1 = \begin{pmatrix} 1 \\ 1 \\ 1 \end{pmatrix}, \qquad a_2 = \begin{pmatrix} -1 \\ 2 \\ 1 \end{pmatrix}, \qquad a_3 = \begin{pmatrix} -3 \\ -2 \\ 1 \end{pmatrix}.$$

Note that the vectors p_i, $i = 1, 2, 3$, indicate the "edges"

of S. If we remove the inequality $a_3' \leq 0$ from the

definition of S, what is the geometrical significance of

p_1, p_2 (defined relative to a_1, a_2, not a_1, a_2, a_3) and z?